Microcosm Manual for Environmental Impact Risk Assessment

Yuhei Inamori
Editor

Microcosm Manual for Environmental Impact Risk Assessment

From Chemicals to Whole Effluent Toxicity (WET)

 Springer

Editor
Yuhei Inamori
Foundation for Advancement of
International Science
Tsukuba, Ibaraki, Japan

ISBN 978-981-13-6800-4   ISBN 978-981-13-6798-4   (eBook)
https://doi.org/10.1007/978-981-13-6798-4

This Springer imprint is published by the registered company Springer Nature Singapore Pte Ltd.
The registered company address is: 152 Beach Road, #21-01/04 Gateway East, Singapore 189721, Singapore

Preface

Chemical substances, such as pesticides and metals, have been regulated through the use of physical and biological containment. However, when trying to regulate field emissions, an extremely troublesome preservation of such chemicals has been encountered in ecosystems in addition to the harmful effects on human health and that of beneficial animals and plants. When incorporating risks to an ecosystem into regulations, a problem arises with respect to determining what to assess. It is impossible to study the reaction of biological communities by releasing chemical substances in nature. This is not limited to chemical substances, as it is also true when assessing the impacts of genetically modified microorganisms on the ecosystems they inhabit or invade. By breeding specific species of *Daphnia*, algae, and fish and from observing and measuring their mortality, growth inhibition, and abnormal behaviors, attempts have been made to evaluate the impact on ecosystems when genetically modified microorganisms or new chemicals are introduced. However, these species are selected for their convenience, such as the ease with which they may be bred, and their observed reactions; they are totally unrelated to the studied ecosystems and therefore may not be used to directly evaluate risks to said ecosystems. Meanwhile, it is possible to implement methods of subdividing the target ecosystem and observing the reactions of local biota to determine the degree of influence of chemicals and genetically modified organisms. However, because in situ experiments lack reproducibility, it is difficult to understand the precise phenomena at work due to the uncertainties associated with subjects in nature.

When targeting an ecosystem for evaluation, it is necessary to know that the following four characteristics are needed for its recognition: "diversity," "heterogeneity," "microscopic," and "reactivity." Additionally, it is difficult to recognize the response of an ecosystem without understanding its "operability." Among these characteristics, diversity and heterogeneity are caused by the open nature of the ecosystem. Next, microscopic characteristics provide a macroscopic impression of an ecosystem, but, in order to see the effects of these characteristics on an ecosystem, it is desirable to consider the changes in the ecosystem, such as through the use of microorganisms as indices. Due to the movement of bacteria, algae and protozoa occupy fundamental positions in the overall function of ecosystems. The reactions of microorganisms are faster than those of animals and plants. Moreover, they respond to pure chemical substances so vigorously that they can record changes in the environment sensitively and promptly.

In order to use microscopic phenomena in an ecosystem to evaluate the influence of chemicals and genetically modified organisms, it is indispensable to manipulate the "expansion" as a matter of course. "Drawing" the ranges of targeted ecosystems remains an important problem. It is impossible to cover the entirety of nature that we perceive. Even though understanding the natural environment is the ultimate purpose of environmental research, it cannot be the actual target of investigation. The subdivision and simplification of ecosystems are inevitable steps that cannot be avoided. However, when trying to simplify environmental components and biota to the scope of an impact assessment, two important problems arise. One is the arbitrariness with which one extracts different kinds of environmental data from natural ecosystems; the second results from having simplified the system to the extent that one loses sight of the connection between the model and the natural environment. We often try to simplify ecosystems by developing mathematical models of them, but real ecosystems not only are composed of diversity but also vary depending upon location and season. Therefore, there is often a non-negligible gap between the model and reality. Indeed, it is unlikely that we can develop a model capable of describing natural diversity in a unified way. As a next measure then, an entity model of a given ecosystem should become a standard. As such, a model reflects the general attributes of the ecosystem; even if there is no structural correspondence with the actual ecosystem, it is reproducible, is rich in operability, responds promptly to external forces, and is easy to measure and recognize. Positioning the phenomena described by such standardized entities as a filter for impact assessment will provide valuable information when incorporating risks to ecosystems into regulatory policies. Research using such information at a given locality would then be the next step.

In this manual, a microcosm test was performed in which microcosms consisting of bacteria as decomposers, micro-animals as consumers, and microalgae as producers were used to evaluate the ecosystem-level effects of chemical agents and wastewater consisting of micropollutants on an aquatic ecosystem. Reproducibility and correspondence to the modeled natural ecosystems proved to be very high in studies on the material cycle, energy flow, and interactions of microorganisms in the microcosm system. The microcosm systems were not comparable with point locations, such as natural lakes, ponds, or rivers, but they were comparable with respect to the function and structure of natural aquatic ecosystems. Based on this work, the above approaches were suggested and applied to predict the effects of chemical pollutants in a natural ecosystem. It was determined that the microcosm test was of great importance when used as a model for biological assessments.

A microcosm using mice was employed as the model system to evaluate the toxicity of chemicals and genetically modified organisms to ecosystems and humans. Estimation from 106 examples of target materials (6 of surfactants, 3 of germicides, 14 of herbicides, 12 of pesticides, 2 of endocrine-disrupting chemicals, 1 of an antibiotic, 1 of an algal bloom toxin, 6 of organic matter and solvents, 2 of nutrients, 15 of metals, 2 of radiation, 6 of microbial pesticides, 8 of genetically modified bacteria, 13 involving biomanipulation, and 2 incorporating climate change; Chap. 7) and 13 examples of the Whole Effluent Toxicity (WET) test (Chap. 9) is described in this book. In the study reported here, research and development into the environmental impact assessment of a chemical substance was conducted, paying close attention to the microcosm, which is an aquatic model ecosystem consisting of producers (phytoplankton), predators (zooplankton), and decomposers (bacteria). Enactment *per* the international guidelines of the Organisation for Economic Co-operation and Development (OECD) test of the general purpose microcosm method in Japan was aimed for by setting the quantity of production/respiration (P/R ratio), which can be used as a measure of change for the entire ecosystem that can be incorporated into an assessment index. Traditionally, environmental impact assessments of chemical substances have been performed using a single species, but model ecosystems that imitate nature may be used to address the problems of stability, reproducibility, and high costs; a standard test method has not been established. However, a foundational manual concerning test operations has already been established (Funds for the Overall Promotion of Environmental Research, Ministry of the Environment, in FY2009-2011), and the ring test, which will be performed in collaboration with several research institutions and developed for wide use, is already planned to establish the OECD standard test method.

This book is composed of the following chapters:

1. Introduction
2. Standardization of the Microcosm N-System
3. Maintenance of the Microcosm N-System
4. Preparation of the Microcosm N-System
5. Procedure for Using the Microcosm N-System
6. Estimation Using the Microcosm N-System
7. Example Assessments of the Microcosm N-System
8. Coefficient of Assessment for the Microcosm System
9. Application to the Whole Effluent Toxicity Test
10. A Scaled-Up Model Ecosystem Verification of the Microcosm N-System

11. A Subsystem Microcosm Verification of the Microcosm N-System
12. Further Perspectives

The superiority of the N-system when analyzing and evaluating the research and development processes and results of microcosm N-system experiments is made clear in this book. A fixed rule of a no-effect density (as the environmental risk) introduced a branching-type analysis of variance into the analytical method, and the case study involved many chemical substances, global warming as the recent environmental problem, acid rain, biomanipulation, and radiation. We hope that this book will provide researchers concerned with environmental problems a system of rating environmental risks to ecosystems and that it will serve as a foundational manual for administrative leaders.

Ibaraki, Japan   Yuhei Inamori

Recommendation

A book recommended for all environmental researchers

It is well known that more than 100,000 types of chemical substances are used by humans, but only a small number of them are regulated, and such regulations vary from country to country. Approximately 50 of them are being used and regulated in Japan. There are physicochemical methods and biological methods for assessing the environmental impacts of chemical substances, but most evaluations are performed using physicochemical methods. Furthermore, physicochemical techniques are not useful for unknown substances and cannot evaluate compounded effects or past effects. On the other hand, biological methods, called biological response tests (i.e., bioassays), present a way to evaluate such impacts by relying on real organisms. Biological response tests on wastewater are often referred to as Whole Effluent Toxicity (WET) tests, and standard test methods have been established in Europe, the United States (US), and Japan, and they may be used in some areas for regulating wastewater.

Because death is the strongest response that organisms may have to chemical substances, chemical concentrations are often shown as half-lethal concentrations

(LC_{50}). Lethal responses appear within a short time, reflecting acute toxicity. Physiological effects on reproduction (proliferation), behavior, teratogenicity, and mutagenicity are felt with chronic poisons. Chronic poisoning is evaluated from the concentration of the effect of exposure to chemical substances over a long period of time by generations, such that abnormalities appear even if an organism (or population) does not die. From this influence, the maximum no-effect concentration (NOEC) is calculated, and the reference value and target value are calculated by multiplying this effect concentration by the uncertainty factor (safety factor). This is the basis of biological response tests performed in laboratory settings.

Organisms used for response testing in Japan are a green alga (*Pseudokirchneriella subcapitata*), a crustacean (*Ceriodaphnia dubia*), and a fish (*Danio rerio*). These were present at the concentration under which chronic toxicity appears. As drainage is discharged into public waters, it is diluted at least ten times over, so we assumed a safety factor of 10. Considering the susceptibility of each species, we assumed that the effect of species differences was 10. It is common to multiply the dilution ratio and the species difference to 100.

However, repeating biological response tests for many organisms using a single organism many times cannot accurately reflect the impact when incorporating interactions among species in ecosystems. The impact evaluation method that partitions natural ecosystems is called "coral," and the experimental method that uses outdoor rivers and ponds to simulate nature is called a "mesocosm," and both of them are often used in ecological impact experiments. These methods are labor-intensive, exhibit low reproducibility, and are extremely difficult to apply to evaluations of targeted effects.

The microcosm covered in this book considers a combination of many species and uses flask-scale experiments, so that many experiments can be performed at the same time. The microorganisms constituting this microcosm include four kinds of bacteria (as decomposers), one kind of protozoan ciliate, two kinds of metazoan rotifers, one type of metazoan oligochaete (as a consumer), and two kinds of green algae and one type of blue-green alga (as producers) in which reproducibility and stability are excellent, and dynamic equilibrium is established among the species. Dynamic equilibrium can be obtained in a short period of time even if transplantation is performed. The effects of chemical substances and foreign microorganisms can be evaluated from the deviations from this dynamic equilibrium. Until now, as the Chairman of the Biological Response Test Study Committee of the Ministry of the Environment of Japan, I have been establishing standards related to biotoxicity tests, such as the WET test. I believe that evaluating the impacts of chemical and microorganismic toxins to ecosystems is extremely important.

Experimental results of the microcosm shown in this book provide many examples of the microbial phases constructed, as mentioned above. Based on this, there is a high possibility of further development of this methodology. Biological response tests must incorporate a mechanism that can evaluate interactions among organisms, and the first choice should be to incorporate the microcosm test. I strongly recommend that environmental researchers read this book, and I sincerely hope to deepen their understanding of ecosystems through the use of this microcosm test.

NPO Institute of Ecological Engineering   Ryuichi Sudo
(E-TEC), Sendai, Miyagi, Japan

Tohoku University, Sendai, Japan
September 17, 2018

Acknowledgments

We greatly thank late Dr. Yasushi Kurihara, emeritus professor of Tohoku University, who developed the microcosm system first in Japan. All of our studies could not start without his original investigation. Also we would like to show special thanks to Dr. Katsura Sugiura, emeritus professor of Sagami Women's University, for his kindful teaching and essential discussion for the assessment method of microcosm examination using P/R ratio through the project of Environment Research and Technology Development Fund of the Environmental Agency of Japan and the Japan Chemical Industry Association (JCIA)'s new Long-range Research Initiative (LRI). Some of the assessment data in this book were cited from Dr. Sugiura's published papers. Furthermore, greatly thanks to Dr. Ryuichi Sudo, former professor of Tohoku University; Dr. Zen-ichiro Kawabata, professor of the Research Institute for Humanity and Nature; Dr. Shuichi Shikano, associate professor of Tohoku University; Dr. Shoichi Fuma, senior researcher of the National Institute for Quantum and Radiological Sciences and Technology; Dr. Takashi Amemiya, professor of Yokohama National University; Dr. Toshiyuki Nakajima, professor of Ehime University; Dr. Tomoaki Itayama, professor of Nagasaki University; and all of the students and associate researchers who investigated microcosm study with us for their kindly cooperation and efforts.

The experimental data described in this book are mainly based on "Development of an Ecosystem Risk Impact Assessment System Using Microcosm (S2-09)" supported by the Environment Research and Technology Development Fund of the Environmental Agency of Japan in the Fiscal years 2009–2011, "Development of an Ecosystem Risk Impact Assessment System Against Chemical Substances Using Microcosm (2012PT4-02)" supported by the Japan Chemical Industry Association (JCIA)'s new Long-range Research Initiative (LRI) in the Fiscal years 2012–2014, and Doctoral thesis of Dr. Kazuhito Murakami (Toho University, 1994), Dr. Nobuyuki Tanaka (Tottori University, 1996), Dr. Yoshie Takamatsu (Tsukuba University, 1999), Dr. Ken-ichi Shibata (Yokohama National University, 2013), and Dr. Kunihiko Kakazu (Fukushima University, 2014).

Contents

About the Editors and Contributors

Editor

Dr. Yuhei Inamori earned his Ph.D. in Biology from Tohoku University, Japan. He joined the Meidensha Co. Ltd., became a senior researcher at the National Institute for Environmental Studies (NIES), and became a professor at Fukushima University. His current research interests include water and waste treatment technologies, bio-eco engineering, energy-saving technologies, and integrated watershed management. Dr. Inamori has also served as President of the Japanese Society of Water Treatment Biology and has received many excellent paper awards from institutions like the Japan Society on Water Environment. He has also been the recipient of the Friendship Award of the Chinese Government, the Minister's Prize, and the Ministry of Environment, Japan.

INAMORI, Yuhei, D.Sc.

Director, Bio-Eco Engineering Research Development Institute, Foundation for Advancement of International Science

President, NPO Bio-Eco Engineering Research Institute
(Preface, Chapters 1, 2, 9, 10, 11, 12)

Contributors

Dr. Ryuhei Inamori received his Ph.D. from Tsukuba University, Japan, and became a senior researcher at Fukushima University. Dr. Inamori has simultaneously served as the Vice President of the NPO Bio-Eco Technology Research Institute. Dr. Inamori's current research interests include water and waste treatment technologies, eutrophication countermeasures, water environment restoration technology, decentralized wastewater treatment technology, energy recovery from organic waste (biomass), environmental education, bio-eco engineering, energy-saving technologies using AOSD (Automatic Oxygen Supply Device) systems and integrated watershed management, and using microcosms for environmental risk assessment. He has received excellent paper awards from the Japan Society on Water Treatment Biology.

INAMORI, Ryuhei, Ph.D.
Associate Director, Bio-Eco Engineering Research Institute, Foundation for Advancement of International Science
Vice President of NPO Bio-Eco Technology Research Institute
(Chapters 1, 9, 10, 11, 12)

Dr. Kunihiko Kakazu received his D.Sc.Tech. in Biology for his research on the Development of Ecosystem Risk Assessment Methods using Microcosms from Fukushima University, Japan. He became a researcher at the Foundation for Advancement of International Science and has been undertaking studies on microcosms and bio-eco engineering.

KAKAZU, Kunihiko, D.Sc.Tech.
Researcher, Foundation for Advancement of International Science
(Chapters 3, 4, 5, 6, 7, 8)

Dr. Yuzuru Kimochi completed his doctoral course in the Department of Agricultural Science at the University of Tsukuba, Japan, and received his Ph.D. with a research theme on wastewater treatment technology that can prevent global warming and eutrophication. As a researcher at the Ibaraki Prefectural Science and Technology Promotion Foundation, he conducted research on wastewater treatment technology. He became a researcher of the Waste Management Group of the Center for Environmental Science in Saitama and conducted research mainly on waste landfill final disposal technology. He became a researcher at the institute's Water Environment Group and is currently a senior researcher. In addition to wastewater treatment technology, he is also undertaking research on biological research techniques that utilize environmental DNA analysis, fish habitat conservation methods, and water quality and environment improvement technology of rivers, among others.

KIMOCHI, Yuzuru, Ph.D.
Senior researcher, Center for Environmental Science in Saitama
(Chapters 9, 10)

Dr. Kazuhito Murakami received his D.Sc. in Biology from Toho University, Japan. He was a researcher at the Okayama Prefectural Institute for Environmental Science and Public Health, and became a research associate of the Chiba Institute of Technology. He has been undertaking study on microcosms and bio-eco engineering. Dr. Murakami also served as President of the Kanto Branch of the Japan Society on Water Environment (JSWE) and as Vice Secretary-General of the Kanto Branch of the Japan Society of Civil Engineering (JSCE).

MURAKAMI, Kazuhito, D.Sc.
Professor/Dean of Advanced Engineering, Chiba Institute of Technology
(Chapters 2, 3, 4, 5, 6, 7, 8, 9, 11, 12, appendix)

Dr. Kakeru Ruike received his Ph.D. in Environmental Studies from the University of Tsukuba, Japan. His doctoral thesis is about the behavior of cyanobacterial toxin in agricultural ecosystems, such as its absorption into edible crops and sorption onto soil particles through irrigation from eutrophic lakes and ponds. He has continued his research into the monitoring and maintenance of this particular strain of microcosm.

RUIKE, Kakeru, Ph.D.
Researcher/Bio-Eco Engineering Research Institute, Foundation for Advancement of International Science
(Chapters 3, 4, 5, 6, 7, 8)

Dr. Ken-ichi Shibata received his Ph.D. in Environmental Science from Yokohama National University, Japan. He served as a research associate at Yokohama National University and Toyo University, undertaking microcosm study. He has been studying oscillations and synchronization in biological systems and green synthesis of metal nano-particles.

SHIBATA, Ken-ichi, Ph.D.
Adjunct teaching staff, Yokohama National University
(Chapters 3, 4, 5, 6, 7, appendix)

Dr. Rie Suzuki received her Ph.D. from Tsukuba University, Japan. Innovative results have been obtained by Dr. Suzuki on the effectiveness of strengthening purification restoration in buffering areas of ecological engineering technology to lower the risk to the environment. This research has been conducted by analyses on the relationship among the purification function stabilization, biological pollution index, and food chain ecosystem. Further, she has undertaken evaluations on on-site domestic wastewater treatment by the Johkasou disposal system. Dr. Suzuki has been a visiting lecturer at Fukushima University and Toyo University and has engaged in risk management of the environmental safety of water, medicine, and food. Dr. Suzuki has received the highest award from a Japan Sewage Works Association presentation session.

SUZUKI, Rie, Ph.D.
Technical Chief Director, Ibaraki Pharmaceutical Association
(Chapter 3)

Dr. Kaiqin Xu received a B.Sc. in Hydro-power Engineering from the Wuhan University of Hydraulic and Electric Engineering (now Wuhan University), China, and an M.S. and a Ph.D. in Civil Engineering (water supply and water treatment) from Tohoku University, Japan, respectively. Dr. Xu held a position as an assistant professor and as an associate professor at the Department of Civil Engineering of Tohoku University. Dr. Xu joined the Water and Soil Environmental Division of NIES, Tsukuba, Japan, as a senior research scientist. He was a section leader in Environmental Restoration and Conservation Technology, NIES. Dr. Xu is interested in the research fields of watershed management, eutrophication countermeasures, water environment restoration technology, decentralized wastewater treatment (rural sewage treatment) technology, energy recovery from organic waste (biomass), environmental education, etc. Dr. Xu has received many excellent paper awards from the Japan Society on Water Environment, Water Treatment Biology, etc.

XU, Kaiqin, Ph.D. (Dr. Eng.)
Professor/Principal Researcher, National Institute for Environmental Studies (NIES), Japan
(Chapters 2, 12)

Chapter 1
Introduction

Yuhei Inamori and Ryuhei Inamori

Abstract The importance of the current conditions of environmental pollution from chemical substances and their environmental impact statement are provided here. The problems of the current test methods (i.e., examinations of single species) and the need for having many kinds of organisms examined are expounded. With this in mind, the summary and effectiveness of the microcosm test are shown and are the main purpose of this book.

1.1 Background

Testing methods for assessing the impact of chemical substances on an ecosystem are divided into single-species and multiple-species tests. Single-species testing has hitherto been widely used for assessing the environmental impact of chemical substances, and various standardized methods have been developed (Beyers and Odum 1993; Graney et al. 1994). However, it is predictable in natural environments that the presence or absence of interactions among different species may give rise to the onset of different toxicity mechanisms caused by chemical substances. The testing of model ecosystems with multiple species accounts for the interactions among different species; hence, it is a robust approach for conducting more realistic risk assessments of ecosystems. Although the need for risk assessment has been acknowledged globally, the development of a standardized testing method has been delayed. Therefore, the promotion of an official, standardized, and generalizable method of testing model ecosystems, with the aim of global applicability, holds great importance.

In Europe and the United States, model ecosystem tests are used in high-risk assessment processes, such as for exposure to pesticides. Test guidelines developed by the Office of Prevention, Pesticides, and Toxic Substances (OPPTS) of the US Environmental Protection Agency (EPA) recommend model ecosystem tests, such as the "OPPTS 850.1900 Generic freshwater microcosm test, laboratory," as methods aimed at understanding the dynamics of chemicals and at measuring their

Y. Inamori (✉) · R. Inamori
Foundation for Advancement of International Science, Tsukuba, Ibaraki, Japan
e-mail: y_inamori@fais.or.jp; r_inamori@fais.or.jp

© Springer Nature Singapore Pte Ltd. 2020

Y. Inamori (ed.), *Microcosm Manual for Environmental Impact Risk Assessment*,
https://doi.org/10.1007/978-981-13-6798-4_1

impact on generic freshwater ecosystems. Additional examples include the "OPPTS 850.1925 Site-specific aquatic microcosm test, laboratory," in which testing is performed by reproducing a specific aquatic ecosystem, and the "OPPTS 835.3180 Sediment/water microcosm biodegradation test," which measures biodegradation in the bottom sediments at a given study site. However, the problem with these approaches is that they lack standardized methods for creating a model ecosystem. Furthermore, the reason model ecosystem testing is less frequently used in the process of evaluating high risks, despite its utility, is that model ecosystem testing is generally costlier than single-species testing (U.S. Environmental Protection Agency 1980, 1981, 1982).

It is important to note that microbial ecosystems, consisting primarily of producers (algae), low-level consumers (microanimals), and decomposers (bacteria), constitute the foundation of aquatic ecosystems. High-level predators such as fish, together with the microbial ecosystem, play an especially important role in water purification and material cycling in aquatic ecosystems. The microbial ecosystem is composed of algae (as photosynthetic primary producers), microscopic animals (acting as consumers), and heterotrophic bacteria (acting as decomposers). It is important to consider the variation in parameters of the ecosystem due to contamination from chemical substances, such as nitrogen, phosphorus, pesticides, and heavy metals. The microcosm method described in this manual is a test that utilizes a flask-scaled model ecosystem that is sampled to form a microcosm, which is equipped with the requirements necessary to solve the aforementioned challenges. An important characteristic of this test is that by designating production (P) and respiration (R) as endpoints in the aquatic microbial ecosystem, which can be easily measured, analyzed, and assessed using a dissolved oxygen (DO) meter, it allows researchers to resolve the issues of complexity and high costs associated with traditional model ecosystem testing.

In assessing and analyzing an ecosystem, it is effective to utilize a complementary dynamic analysis of a microbial community that uses a microcosm (i.e., a stable model ecosystem), which has been established based upon aquatic monitoring data and constitutes the core of the microcosm testing. Currently, traditional methods that utilize a single species do not include the performance of ecosystem risk assessments that examine the effects of chemical substances on ecosystem functions. In the single-species techniques that have been used, ecosystem risk evaluations that include the influence of chemical substances on ecosystem functioning have not conventionally been performed. Furthermore, current ecological studies that utilize a microcosm also emphasize the need for official methods (i.e., international standardization) to assess environmental impacts, which can thus be generalized and used to ameliorate the complexity and high costs of assessment methods involved in model ecosystem testing. The Organisation for Economic Co-operation and Development (OECD) test guidelines also discuss the importance of ecosystem assessments. From such a point of view, the development of an ecosystem-scale evaluation of a microcosm model system can be deemed essential.

1.2   Outline of the Microcosm

The term microcosm is derived from the Greek words for "small" (mikrós) and "universe" (kósmos); it denotes a system in which a population of a single species or a community, a group of populations of two or more species, is cultured in a container under controlled conditions. A large number of microcosms have hitherto been created to elucidate microbial interactions and their mechanisms, as well as to assess the effects of hazardous chemical substances and foreign microbes on ecosystems from the perspectives of microbial ecology and environmental science. These microcosms are classified into three types by size: (1) real-world scale (mesocosm), (2) pilot plant (pilot site) scale, and (3) flask scale. Furthermore, they can be divided into three types according to their population compositions: (1) gnotobiotic, in which the species composition is fully known, the population sizes of each species can be measured, and the traits of each species can be analyzed in isolation; (2) stress-selected, in which a natural community is cultured under specific conditions to promote natural selection in an effort to maintain and develop a specific biological community; and (3) naturally derived, in which a real-world community is maintained without varying any conditions. Among these, the concept of the gnotobiotic and the stress-selected types is used in the same sense as a standardized aquatic microcosm and abstract model ecosystem, respectively. Because the abstract model ecosystem (i.e., stress-selected microcosm) can be steadily sustained with repeated subculturing, it is well-suited to repetitive experiments and has been used in both theoretical ecology and applied ecology. It has also recently begun to be used as a test for assessing environmental impacts. The microcosm in our research model consists of producers, consumers, and decomposers. It can be considered to fall under the abstract model ecosystem category with respect to its properties and under the standardized aquatic microcosm category with respect to its structure.

The microcosm in this experiment is not merely a system for microbial cultures. The system is characterized by its ability to replicate the target phenomena at an ecosystem level, as it includes the physical, chemical, and biological factors of an ecosystem and some of their interactions, which substantiate four relationships: proliferation, consumption, production, and inhibition. Therefore, when applying outcomes obtained through a simple experimental system in a laboratory to the interpretation of real-world phenomena, the phenomena observed in a microcosm are expected to play an intermediary role, linking laboratory and natural conditions and responses.

This manual presents microcosms (N-systems) that were originally developed from water in a natural environment at Tohoku University, Japan, by Prof. Yasushi Kurihara (1926–2005) through the process of natural selection and that were subsequently followed and modified by the National Institute for Environmental Studies (NIES), Japan. They were subcultured as stable ecosystems at the Bio-Eco Engineering Research Institute in the Foundation for Advancement of International Science (FAIS), the Chiba Institute of Technology, and the Yokohama National University, Japan. These microcosms are aquatic, microbial model ecosystems that are composed of at least four bacterial species normally observed in natural ecosystems as decomposers, including *Pseudomonas putida*, *Bacillus cereus*, *Acinetobacter* sp., and coryneform bacteria. The consumers include a protozoan ciliate (*Cyclidium glaucoma*), metazoan rotifers (*Philodina erythrophthalma* and *Lecane* sp.), and a metazoan oligochaete (*Aeolosoma hemprichi*), and the producers include green algae (*Chlorella* sp. and *Scenedesmus quadricauda*) and filamentous cyanobacteria (*Tolypothrix* sp.). These microcosms are considered highly reproducible and stable aquatic model ecosystems. When the microcosms are transferred to a new medium during their stable phase, similar proliferation curves are observed, and, once the system reaches a steady state, it will endure for an extended period of time with the same amount of biomass. Therefore, unlike model ecosystems (mesocosms) established from environmental water, this microcosm will not result in the loss of species—the producers, consumers, and decomposers that constitute an ecosystem—and their impact can be properly assessed from the perspectives of function and structure. Thus, this microcosm is an abstract model ecosystem consisting of producers, consumers, and decomposers. Additionally, it has been demonstrated that the system still endures even if a small fish, such as a guppy (*Poecilia reticulata*), is introduced as a high-level predator. Moreover, despite variations in the number of days required to reach a steady state, this microcosm developed into systems with similar species compositions and similar amounts of biomass at various culture temperatures, ranging from 10, 20, 25, and 30 °C. For this reason, using microcosms with differing culture temperatures allows researchers to evaluate the effects of variations in water temperature on ecosystems. Furthermore, with regard to the effects of cesium radiation on ecosystems, in our joint research with the National Institute of Radiological Sciences (NIRS), Japan, we have reported the novel finding that bacteria, algae, protozoans, and metazoans that underlie the food chain are not affected by even high doses of cesium, which supports the feasibility of assessing the impact of different chemical substances.

As discussed above, the microcosm, with its high reproducibility and stability, allows for different approaches in assessing ecosystems from a functional perspective. When viewed as a standardized model for multi-species testing of the impacts of chemical substances and microorganisms on an ecosystem, it is a very effective model. Additionally, it holds a great value as an ecosystem impact test that assesses the effects of chemical substances and microorganisms on the stability of a system in which material cycles and energy flows exist, which are the foundations of any ecosystem (Fig. 1.1).

1.3 Purpose

The aim of the testing method presented in this book is to perform an aquatic ecosystem risk assessment using a microcosm. Various interactions in the ecosystem were exposed to a chemical substance, which served as a pollutant, and these are shown in Fig. 1.2. The microcosm is a model ecosystem that simulates, at a reduced

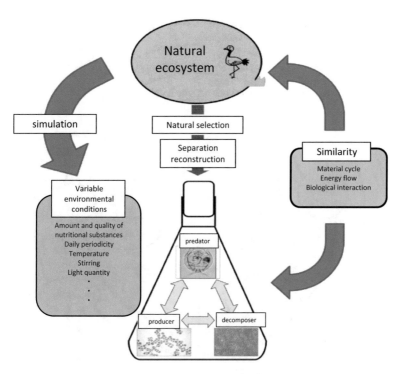

Fig. 1.1 Relationship between natural ecosystem and microcosm system

scale, an aquatic ecosystem that is spatially controllable, and joins laboratory testing and outdoor monitoring (Fig. 1.3). Material circulation and the flow of energy among a variety of organisms are thought to be important in performing an impact assessment of a given ecosystem, which is why there is a limit to rating systems that consist of single species, such as the alga, fish, and water fleas used for ecosystem assessments in the current Organisation for Economic Co-operation and Development (OECD) test. The microcosm (N-system) in question is based upon microbial samples collected from rivers and lakes around Japan by the late Yasushi Kurihara, a Professor Emeritus at Tohoku University. He subcultured samples in a Taub-Peptone (TP) medium (described below) over an extended period of time and confirmed the formation of a stable ecosystem before isolating single microbes and recombining them again as a model system for stable, aquatic ecosystems. Following transfer to the NIES, it was named the "N-system" after the acronym "NIES." In the microcosm, producers (algae), predators (microscopic animals), and decomposers (bacteria) exist as tools for assessing the impact of chemical substances on an aquatic ecosystem. The system encompasses the rules and principles of a functional ecosystem, such as microbial interactions, material cycles, and energy flows, which are not found in single-species culture systems.

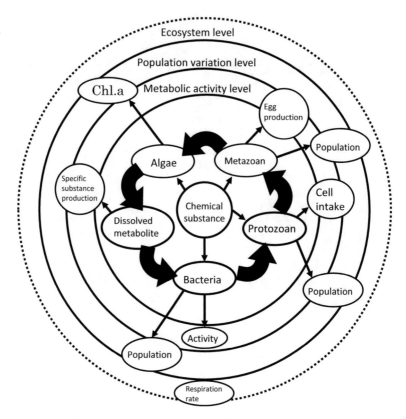

Fig. 1.2 Interaction of pollutants on ecosystem

The N-system is a microcosm that consists of a combination of at least four dominant bacterial species acting as decomposers, four microanimal species acting as predators, and three photoautotrophic species acting as producers. The N-system is a model microbial ecosystem with high stability and reproducibility that, when transferred to a new medium as a seed, during its stable phase, will repeatedly yield a coexisting and coevolving system with a similar proliferation curve to the original microcosm. With over 40 years of stable transfers, it is highly effective as a unified standard for comparing and analyzing data (Fig. 1.4). Furthermore, operation under various conditions is possible while retaining the basic elements of ecosystem functions. Given these characteristics, this model is intended to assess ecosystem risks using production volume/respiration volume (P/R) ratios concomitant with shifts in the entire ecosystem as an indicator. Using the flask microcosm N-system as a model of ecosystem functions, our aim is to offer researchers internationally applicable guidelines for this general microcosm testing method developed in Japan.

Because it is expected that the microcosm is highly correlated with the corresponding natural ecosystem, it is assumed that a more realistic and predictable

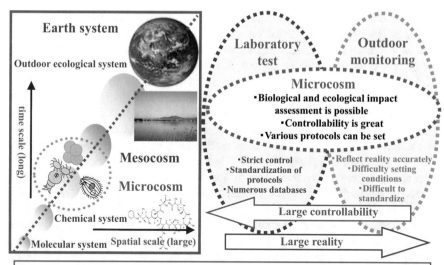

The microcosm system is a time and space controllable sub-ecological system that connects laboratory tests and outdoor monitoring.

Fig. 1.3 Certainty of microcosm test method from hierarchy of nature

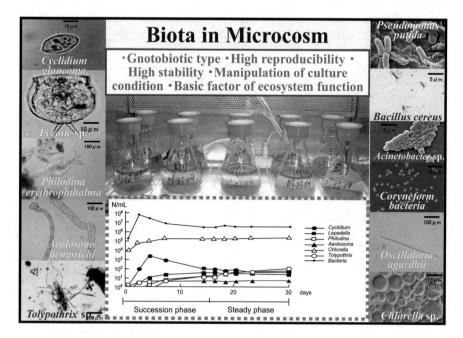

Fig. 1.4 Microorganisms in microcosm and their growth curve

no-effect concentration can be obtained, as compared to the methods currently available (i.e., assessment based on a single species). In assessing the effects of chemical substances on ecosystems, it is difficult to avoid the current testing methods for algae, crustaceans, and fish, which are exemplified in the Whole Effluent Toxicity (WET) test. Using an ecosystem model that includes parallel food chains and energy flows allows us to accumulate knowledge on the decomposition and persistence of chemical substances and on the recovery and disruption of associated ecosystem functions. In short, it is expected that the advantage of microcosms will be appreciated when establishing an approach that numerically assesses ecosystem impacts. The WET test assesses toxicity, including complex effects, by testing water that contains multiple chemical substances rather than assessing the toxicity of each chemical substance in isolation. Although species located in different niches within a food chain are used for the assay, it is a single-species test, and, as previously reported, a drawback of WET testing is that it is conducted under conditions in which material cycles, energy flows, and interactions among different species—the basic components of an ecosystem—are all absent. Microcosms are systems in which multiple species coexist, allowing researchers to assess the risk of chemical substances at the ecosystem level, and the safety coefficients obtained are considered different from those obtained from conventional approaches, as shown in previous studies on the correlations in both mesocosm and microcosm tests. It is expected that further accumulation of data will allow for the calculation of realistic levels of no-effect concentrations predicted for natural ecosystems. The similarity of the P/R ratio between the natural ecosystem and the microcosm is shown in Fig. 1.5. Additionally, the idea underlying the development

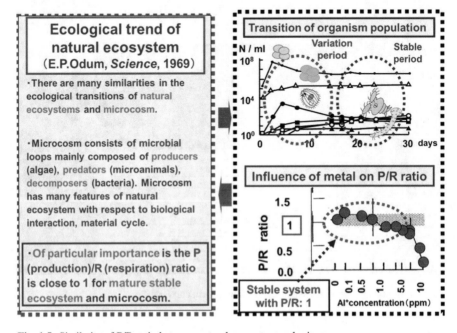

Fig. 1.5 Similarity of P/R ratio between natural ecosystem and microcosm

of the environment risk evaluation technique using the microcosm is illustrated in
Fig. 1.6, which also shows the principles and experimental and analytical methods of
measuring the P/R ratio in a microcosm. The aim of development of environmental
impact risk assessment method using microcosm system is shown in Fig. 1.7.

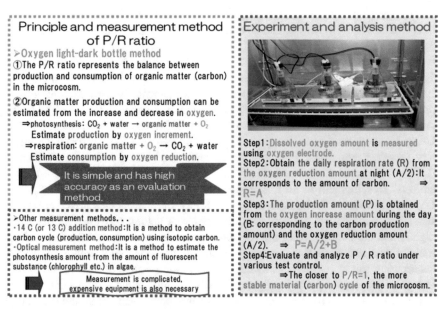

Fig. 1.6 Principle, experiment, and analytical method of P/R ratio in microcosm

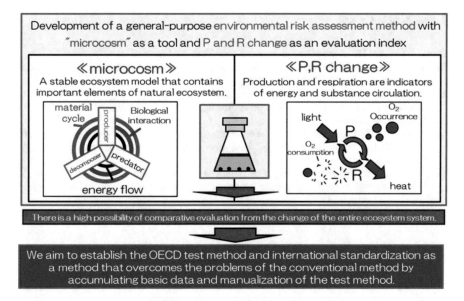

Fig. 1.7 Development of environmental impact risk assessment method using microcosm system

The concept of microcosm testing was discussed in a study entitled "Basic Examination for Discussion on an Outdoor Testing Approach Associated with Usage of Recombinants in Open Systems," which was commissioned by the Planning and Coordination Bureau, Ministry of the Environment of Japan, in the 1989 fiscal year (FY), and also in a study conducted from 1992 to 1993 entitled "Research on the Development of a New Approach for Water Quality Assessment Using Microbial Ecosystem Microcosm (04650505)" supported by the Grants-in-Aid for Scientific Research on Priority Areas (General Research C). On these bases, microcosm testing was listed in the 1997 Sewage Examination Method (Japan Sewage Works Association, Volume III: Biological Examination, Chapter 1: Biological Examination, Section 10: Ecosystem Impact Assessment Testing), and the testing procedures were described in the study conducted during the 2009–2011 FYs in the project entitled "Development of an Ecosystem Risk Impact Assessment System Using Microcosm (S2-09)" supported by the Environment Research and Technology Development Fund. Furthermore, the new Long-Range Research Initiative (LRI) of the Japan Chemical Industry Association (JCIA) in the 2012–2014 FYs, entitled "Development of an Ecosystem Risk Impact Assessment System Against Chemical Substances Using Microcosm (2012PT4-02)," assessed a broader range of substances and conducted correlation analyses with natural ecosystems. The JCIA highlighted the importance of the relationship of microcosm tests with the existing toxicity data and created a database, which sought to define optimum values for determining safety coefficients. Moreover, during the course of testing across different facilities for the purpose of generalization, a ring test that is in line with OECD test guidelines was established and has been subsequently enhanced and revised.

Literature Cited

Beyers RJ, Odum HT. Ecological microcosms. New York: Springer; 1993. 557pp.

Graney RL, Kennedy JH, Rodjers JH Jr. Aquatic mesocosm studies in ecological risk assessment. Boca Raton: Lewis Publishers; 1994. 723pp.

U.S. Environmental Protection Agency. Microcosms as potential screening tools for evaluating transport and effects of toxic substances; Athens, Athens, U.S. Environmental Protection Agency; 1980. 370pp.

U.S. Environmental Protection Agency. Microcosms as test systems for the ecological effects of toxic substances; an appraisal with cadmium, Athens, U.S. Environmental Protection Agency; 1981. 171pp.

U.S. Environmental Protection Agency. Community structure, nutrient dynamics, and the degradation of diethyl phthalate in aquatic laboratory microcosms; Athens, U.S. Environmental Protection Agency; 1982. 135pp.

Chapter 2
Standardization of the Microcosm N-System

Kazuhito Murakami, Yuhei Inamori, and Kaiqin Xu

Abstract In this chapter, the perspective needed for examining graded impact statements is described. Such an evaluation is comprised of the examinations of various culture conditions for standardizing the microcosm N-system, the two-species cultures, the microcosm, and the mesocosm.

2.1 Standardization of the Microcosm N-System

In order to use the microcosm for environmental impact assessments, it is necessary to standardize the microcosm culture conditions. To this end, important parameters of the culture conditions for microbial cultivation were investigated.

2.1.1 Species Composition

Seventeen species, consisting of six species of protozoans and metazoans as predators, seven species of algae as producers, and four species of bacteria as decomposers, were used to investigate the effect of species composition on the stability and reproducibility of the microcosm system, with the aim of developing a standard, aquatic, flask-sized microcosm system. Three species each of protozoans and metazoans were used as predators (consumer). The protozoans (Ciliata) were *Cyclidium glaucoma*, *Tetrahymena pyriformis*, and *Colpidium campylum*. The

K. Murakami (✉)
Chiba Institute of Technology, Narashino, Chiba, Japan
e-mail: kazuhito.murakami@p.chibakoudai.jp

Y. Inamori
Foundation for Advancement of International Science, Tsukuba, Ibaraki, Japan
e-mail: y_inamori@fais.or.jp

K. Xu
National Institute for Environmental Studies, Tsukuba, Ibaraki, Japan
e-mail: joexu@nies.go.jp

© Springer Nature Singapore Pte Ltd. 2020
Y. Inamori (ed.), *Microcosm Manual for Environmental Impact Risk Assessment*,
https://doi.org/10.1007/978-981-13-6798-4_2

metazoans were of two types, namely, rotifers and aquatic oligochaetes. The former included *Philodina erythrophthalma* and *Lecane* sp., and the latter was *Aeolosoma hemprichi*. These protozoans and metazoans frequently appear in both natural ecosystems, such as lakes, marshes, rivers, and seas, and in artificial ecosystems, such as biofilms and activated sludge, and they are easily cultured and accurately counted. These strains were isolated from water treatment facilities and a polluted lake.

Three species of algae belonging to Chlorophyta, *Chlorella* sp., *Scenedesmus quadricauda*, and *Chlamydomonas monticola* and four species of blue-green algae, *Tolypothrix* sp., *Oscillatoria agardhii*, *Anabaena flos-aquae*, and *Microcystis viridis*, were used as producers. The strains were sourced from the Microbial Culture Collection Center of NIES in Tsukuba, Japan. These algal species are often observed in lakes and ordinal ponds. Four species of bacteria were used as decomposers, namely, *Pseudomonas putida* (gram-negative aerobic rod bacteria), *Bacillus cereus* (gram-positive aerobic endospore-forming rod bacteria), *Acinetobacter* sp. (gram-negative rod bacteria), and a coryneform bacterium (gram-positive rod bacteria). Mixed cultures of these bacteria were utilized.

Taub's basal medium (200 mL) containing 100 mg/L of polypeptone was placed in a cotton-plugged 300 mL volume Erlenmeyer flask and sterilized by an autoclave at 1 atm and 121 °C for approximately 20 min. A stock culture of each microorganism was added, and the flask was placed in a growth chamber at 25 °C under cool white fluorescent lighting for 12 h (2500 lux) and in the dark for 12 h, sequentially. Cultivation was conducted under static conditions.

Certain species of protozoan predators were able to be established within the system, but others were not. All species of metazoan predator were able to coexist, and filter feeders, such as rotifers, and detritus feeders, such as oligochaete, were also able to coexist. All species of blue-green algae (as producers) were able to establish themselves, notwithstanding any other microorganisms. All species of chlorophytes were able to be established, and the combination of and interaction between chlorophytes and protozoans appeared to be of importance. All species of bacterial decomposers were able to establish themselves, with none disappearing in any of the microcosms. It was apparent that ecosystem stability did not always increase as a function of species richness. From these results, the species composition of a standardized microcosm was determined as the combination of one species of protozoa, *Cyclidium glaucoma* (Ciliata); three metazoan species, including two rotifers, *Lecane* sp. and *Philodina erythrophthalma*, and one oligochaete, *Aeolosoma hemprichi*, as predators (consumers); two species of chlorophytes, *Chlorella* sp. and *Scenedesmus quadricauda*, and one species of blue-green algae, *Tolypothrix* sp., as producers; and four species of bacteria, *Pseudomonas putida*, *Bacillus cereus*, *Acinetobacter* sp., and a coryneform bacterium.

2.1.2 Culture Vessel

Constructing optimum culture conditions is necessary for generating a standardized microcosm. Specifically, the following must be optimized: (i) the movement of the experimental water body, (ii) the wall effects, (iii) species composition and abundance, and (iv) light and temperature conditions. To ensure an optimum microcosm cultivation volume, an 18 mL test tube and 100 mL, 300 mL, 500 mL, and 1000 mL Erlenmeyer flasks were used for investigation.

In culture vessels of any volume, all microorganisms in the microcosm transited the same growth curve, and no difference was observed among any of the culture vessels. The numbers (N) of each microorganism in the steady state were approximately as follows: 100 N/mL of *Cyclidium glaucoma*, 30 N/mL of *Lecane* sp., 40 N/mL of *Philodina erythrophthalma*, and 10 N/mL of *Aeolosoma hemprichi*. This indicates that the volume of the microcosm, ranging from 18 mL to 1000 mL, does not affect the succession of microbiota for the purposes of conducting the microcosm test. Although there is no influence on the succession of microbiota, a 300 mL Erlenmeyer flask is appropriate from the standpoints of handling, cultivation space, and sampling of the microorganisms.

2.1.3 Stirring

The effect of stirring on the stability of the microcosm was investigated. The microbiota in the microcosm was divided into two groups, influenced species and uninfluenced species, in the succession period. Specifically, the blue-green alga, *Tolypothrix* sp., grew in a fragmentary manner when stirred. However, there was no difference in the microbiota between the stirred microcosm and static microcosm by the 16th day after cultivation began. This indicates that stirring, as an outside factor, disturbs the stability of the microcosm during the succession period but has no affect in the stable period. In the microcosm, which is used as a model of natural ecosystems, spatial inhomogeneity is important as an ecological characteristic. Although some differences were observed in the growth patterns of certain microbiota, a stable system could be obtained under both stirring and static conditions.

2.1.4 Temperature

Culture temperatures were set to 10, 20, 25, and 30 degrees Celsius (°C), and the growth patterns of microorganisms in the microcosm were observed. The effect of temperature on microorganism growth was conspicuous, and the speed of succession was faster under higher-temperature culture conditions. Namely, growth was delayed, the peak abundance of *Cyclidium glaucoma* appeared 20 days after cultivation began, and a high population density was maintained at 10 °C. However, the population sizes of other microanimals were lower than 10 N/mL, and the dominant species of bacteria was represented by only three individuals. At 20 °C, the peak abundances of bacteria and *Cyclidium glaucoma* appeared 7 days after cultivation began, and the population of *Cyclidium glaucoma* stabilized at ~150 N/mL. Other microorganisms also had more than 1 N/mL after 21 days, but the dominant species of bacteria was represented by only one individual. At 25 °C, all microorganisms exhibited concentrations of more than 1 N/mL by the 7th day, and a steady state was maintained after the 14th day. In the steady state, microanimals coexisted as ~100 N/mL of *Cyclidium glaucoma*, 30 N/mL of *Lecane* sp., 40 N/mL of *Philodina erythrophthalma*, and 10 N/mL of *Aeolosoma hemprichi*. At 30 °C, the succession pattern was fairly similar to that at 25 °C, but a decrease in the abundance of *Cyclidium glaucoma* by the 7th day was conspicuous. The microcosm culture allowing for high stability and serial transferring on a monthly cycle was determined to be 25 °C.

2.1.5 Illuminance

Material circulation in ecosystems begins with photosynthesis by plants and microalgae, such as chlorophytes and blue-green algae, which filled this niche in the experimental microcosm. To establish a standard condition for illuminance, the light/dark (L/D) cycle was investigated. An L/D cycle, with 12 hr each, was considered appropriate because the succession of microbiota was smooth. Even under a 16 hr/8 hr cycle of light/dark illuminance, the microcosm reached a steady state; however, the lifespan of the fluorescent lamp was short. Under both 24 hr/0 hr (total light) and 0 hr/24 hr (total darkness) conditions of illuminance, the succession of microbiota was unstable.

2.1.6 Substrate Concentration

While the medium is important for cultivating the microcosm, the substrate concentration is of greater importance. To obtain a stable microcosm, the effect of the substrate concentration was investigated; 25 mg/L, 50 mg/L, and 100 mg/L of polypeptone were added to Taub's basal medium, and the microcosm was cultured under these conditions. The microcosm was able to reach a steady state under all of these conditions, but there was an observed tendency toward population decline and extinction of some species of microorganisms under the 25 mg/L of polypeptone condition. From the results obtained in these experiments, it was determined that a 100 mg/L concentration of polypeptone was safe.

2.1.7 Standard Cultivation Conditions of the Microcosm N-System

From the experiments described in Sects. 2.1.1, 2.1.2, 2.1.3, 2.1.4, 2.1.5, and 2.1.6 of this chapter, the standard culture conditions of the flask-sized microcosm were determined as follows: (i) species composition involves the combination of one species of ciliate protozoan, *Cyclidium glaucoma*; three species of metazoan predators, two rotifers, *Lecane* sp. and *Philodina erythrophthalma*, and one oligochaete, *Aeolosoma hemprichi*; two species of chlorophytes, *Chlorella* sp. and *Scenedesmus quadricauda*; and one species of blue-green algae, *Tolypothrix* sp., as producers, and four species of bacteria, *Pseudomonas putida*, *Bacillus cereus*, *Acinetobacter* sp., and a coryneform bacterium; (ii) the culture vessel size is that of a 300 mL Erlenmeyer flask; (iii) there is no stirring; (iv) culturing occurs at 25 °C and under 2400 lux, with a 12 hr/12 hr L/D cycle; (v) Taub's basal medium is supplied with 100 mg/L of polypeptone as the substrate. A basic pattern that was observed was an increase in the abundance of microorganisms in the microcosm under the standard culture conditions (Fig. 2.1). A similar increase was observed when a new nutrient medium was provided to seed the microcosm N-system, continuing the transition toward the stationary phase. In other words, the microcosm N-system has very high plasticity and stability, and the system is thus a superior tool for repeated experimentation.

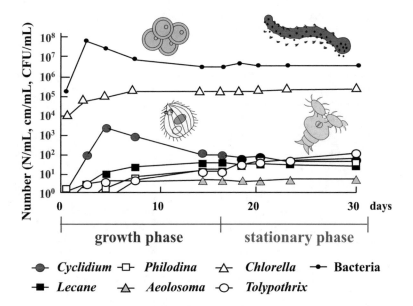

Fig. 2.1 Growth pattern of microorganisms in microcosm N-system

2.2 Environmental Assessment Using a Model Ecosystem

To estimate the effect of chemical substances on the microcosm, the step-by-step test method, consisting of three tests, was considered to be essential (Japan Sewage Works Association 2012). A conceptual diagram of the examination of the graded ecosystem impact is shown in Fig. 2.2. A two-species culture test was used to investigate prey-predator interactions between bacteria and microanimal predators. Microanimal predators were selected from among the protozoans and metazoans, taking note of the frequency of their appearances in both natural ecosystems, such as rivers, lakes, marshes, and seas, and in artificial aquatic ecosystems, such as biofilms and activated sludge, as well as the ease and accuracy with which they may be cultured and counted. The microanimals used were *Cyclidium glaucoma*, *Tetrahymena pyriformis*, *Colpidium campylum*, *Philodina erythrophthalma*, and *Aeolosoma hemprichi*. They have been cultured and serially transferred in a lettuce and egg (LE) medium for several years.

The flask microcosm described in this book consisted of four species of bacterial decomposers, one species of protozoa, three species of metazoan predators (consumers), and three species of microalgae as producers. This system displayed very high reproducibility and a robust reflection of a natural ecosystem. It is therefore suggested that this small-scale, repeatable microcosm can be used as a tool for screening tests at the ecosystem level.

A natural lake model ecosystem (i.e., a mesocosm) constitutes natural lake water in which naturally occurring microbiota exist (Graney et al. 1994; Harris 2013). It contains several species of predatory microanimals (consumers), microalgae as producers, and bacterial decomposers. This system can serve as an intermediate

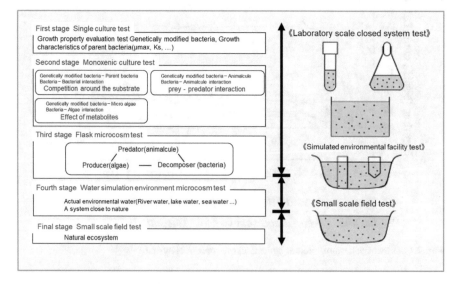

Fig. 2.2 Conceptual diagram of stepwise ecological impact assessment test (tier test)

between a flask-sized microcosm and a natural aquatic ecosystem. As described above, using a step-by-step assessment method (i.e., a tier test), the safety of target chemicals, genetically engineered microorganisms, and microbial pesticides can be more accurately evaluated.

Literature Cited

Beyers RJ, Odum HT. Ecological microcosms. New York: Springer; 1993. 557pp.

Graney RL, Kennedy JH, Rodjers JH Jr. Aquatic mesocosm studies in ecological risk assessment. Boca Raton: Lewis Publishers; 1994. 723pp.

Harris CC. Microcosms; ecology, biological implications and environmental impact. New York: Nova Publishers; 2013. 189pp.

Japan Sewage Works Association. Wastewater Examination Method -FY2012-; Tokyo, 2012.

Tanaka N, Inamori Y, Murakami K, Akamatsu T, Kurihara Y. Effect of Species Composition on Stability and Reproductivity of Small-scale Microcosm System. Water Sci Technol, 1994;30 Suppl 10:125–131.

Chapter 3
Maintenance of the Microcosm N-System

Ken-ichi Shibata, Kunihiko Kakazu, Kakeru Ruike, Kazuhito Murakami,
and Rie Suzuki

Abstract In this chapter, the procedures for maintaining and managing a standard microcosm (N-system) and the cultivar installation and cultivation methods introduced in this book are described. The characteristics of microorganisms in the microcosm are also described.

3.1 Apparatus

- 300 mL Erlenmeyer flask
- Silicon plug (Matsuura Machinery, Catalog No. 6-343-16, Type No. C-40)
- Incubator (25 °C; light/dark (L/D) cycle = 12 hr/12 hr; illuminance = 2400 lux (photosynthesis photon flux density = 36 μmol/m^2/s))

3.2 Cultivation

The maintenance culture of the microcosm N-system must be aseptic. A TP (Taub + peptone) medium (shown in Table 3.1) was used for cultivation. After adjustment of the TP medium, 300 mL was dispensed into an Erlenmeyer flask by 200 mL and

K. Shibata
Yokohama National University, Yokohama, Kanagawa, Japan
e-mail: shibata-kenichi-ym@ynu.jp

K. Kakazu · K. Ruike
University of Tsukuba/Foundation for Advancement of International Science, Tsukuba, Ibaraki, Japan
e-mail: kakazu@fais.or.jp; ruike@fais.or.jp

K. Murakami (✉)
Chiba Institute of Technology, Narashino, Chiba, Japan
e-mail: kazuhito.murakami@p.chibakoudai.jp

R. Suzuki
Ibaraki Pharmaceutical Association, Mito, Ibaraki, Japan
e-mail: suzuki.r@ibaraki-kensa.or.jp

© Springer Nature Singapore Pte Ltd. 2020
Y. Inamori (ed.), *Microcosm Manual for Environmental Impact Risk Assessment*,
https://doi.org/10.1007/978-981-13-6798-4_3

Table 3.1 Taub Polypeptone medium

Stock solutions				Volume
A	Polypeptone*[2, 3, 4]	2.0 g	/100 ml	10.0 ml
B	$MgSO_4 \cdot 7H_2O$	4.93 g	/200 ml	2.0 ml
C	KH_2PO_4	2.72 g	/200 ml	2.0 ml
	NaOH	0.56 g		
D	$CaCl_2 \cdot 2H_2O$	7.35 g	/500 ml	40.0 ml
E	NaCl	2.92 g	/500 ml	60.0 ml
F	$FeSO_4 \cdot 7H_2O$	4.98 g	/200 ml	0.375 ml
	Na_2 EDTA	2.72 g		
G	H_3BO_4	0.930 g	/500 ml	1.0 ml
	$ZnSO_4 \cdot 7H_2O$	0.144 g		
	$MnCl_2 \cdot 4H_2O$	0.99 g		
	$Na_2MoO_4 \cdot 2H_2O$	0.012 g		
	$CuSO_4 \cdot 5H_2O$	0.025 g		
	$Co(NO_3)_26H_2O$	0.145 g		
D.W.				Up to 2000 ml

Total nitrogen (T-N) = 100 mg/L × 14.5% = 14.5 mg/L
Total phosphorus (T-P) = 1.0(peptone) + 3.1(KH_2PO_4) = 4.1 mg/L
T-N/T-P = 3.54

capped with a silicon plug; it was then autoclaved at 121 °C for 15 min to sterilize it. After having cooled to room temperature, 30 mL of the N-system, which was cultured for 2–8 weeks, was added as a seed and cultured with an incubation involving an L/D cycle = 12 hr/12 hr, with an illuminance of 2400 lux (photosynthesis photon flux density = 36 μmol/m²/s) at 25 °C. The culture was stirred lightly by hand once *per* day until culturing began 1–3 days later, after which time it was allowed to rest. Subculturing was performed once every 2 months. A National (a.k. a., Panasonic) FL20SS-W/18 was used as a source of light, and all fluorescent bulbs were changed every 3 months or as needed based upon the illumination or a photosynthesis photon flux density, measured every month. The conditions of incubation of the microcosm (Figs. 3.1 and 3.2) (i.e., L/D = 12 hr/12 hr, illuminance = 2400 lux, and 25 °C) as well as the appearance (i.e., the culturing N-system in a 300 mL volume flask capped with a silicon plug) of the microcosm are shown in Fig. 3.2. Measurement of production and respiration amount is shown in Fig. 3.3.

- Silicon plug: Matsuura Machinery, Catalog No. 6-343-16, Type No. C-40; ventilation amount = 2000–4500 mL/min.
- Polypeptone: Wako Chemicals Co., Code No.394-00115 (500 g), 390-00117 (10 kg). Polypeptone (Nihon Pharmaceutical Co., Ltd.) is an ashy yellow dry powder, which is purified after zymolysis of milk casein, and is equivalent to the Trypticase Peptone and Casitone of BD Co., Ltd.
- Additional methods of Polypeptone depended upon each experiment (e.g., it was added as a powder each time or an autoclaved stock solution of Polypeptone was made and added aseptically each time, etc.).

Fig. 3.1 Microcosm
N-system

Fig. 3.2 Incubator

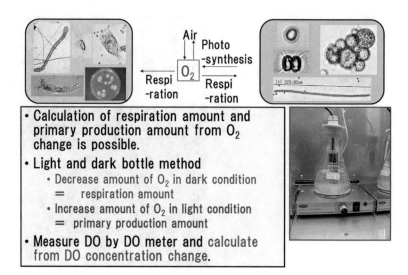

Fig. 3.3 Measurement of production and respiration amount

- The nitrogen content of Polypeptone is 14.5% (w/w), and the phosphorus content is 1.0% (w/w), as measured by the Japanese Industrial Standard (JIS) K-0102 method.

3.3 Microorganisms in the Microcosm N-System

The Kurihara-type microcosm N-system is a gnotobiotic microcosm in which the species composition is completely known; experimentally, this is superior to the stress-selected type of microcosm, in which environmental water is added to seed the nutrient medium and culture. The naturally derived type is one in which the culture is derived from the environmental water sampled from a natural lake or marsh in equilibrium with and representative of the system that is necessary for impact assessment. The microcosm N-system is a multi-species culturing ecosystem model and consists of two species of Chlorophyceae, *Chlorella* sp. and *Scenedesmus quadricauda*, and one species of cyanobacteria, *Tolypothrix* sp., as producers; one species of ciliate protozoan (*Cyclidium glaucoma*), two species of rotifers (*Lecane* sp. and *Philodina erythrophthalma*), and one species of oligochaete (*Aeolosoma hemprichi*) as consumers (predators); and four species of bacteria, *Bacillus cereus*, *Pseudomonas putida*, *Acinetobacter* sp., and a coryneform bacterium, as decomposers. The biological characteristics of each microorganism are described in Sects. 3.3.1, 3.3.2, and 3.3.3 below (Tanaka 2002; Sudo and Inamori 1997; Shiga Scientific Teaching Materials Research Committee 2005; Japan Sewage Works Association 2012; Holt 1994).

3.3.1 Microalgal Producers

Chlorella sp.

The *Chlorella* sp. in the microcosm is a free-living (it remains uncertain if they float or attach to a substrate during ontogeny), single-cell alga related to the Chlorophyceae. The cells have a chloroplast and pyrenoid within a globe, and the diameter of each cell is 5–10 μm. *Chlorella* sp. form cohesive bodies (i.e., flocculate) in the microcosm, and a single individual may be as large as a formed floc. Population measurements assume that each individual is equal to one cell, regardless of the size of the cell.

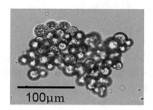

Scenedesmus quadricauda

Plural cells of *Scenedesmus quadricauda* form a fixed-number colony, which affect one line, and are free-living (it is uncertain if they float or attach to a substrate), and the cells have an oval chloroplast. Colonies of four to eight cells are frequently observed. An individual single cell, which does not form a colony, and colonies of two cells rarely appeared in the microcosm. One colony is counted as one individual, regardless of the number of individual cells. In recent years, species that exhibit a typical shape, with a cell with bow-shaped spines at both ends of the colony, have been revised as a species of *Desmodesmus* (*Desmodesmus* sp.). However, in the microcosm, *Scenedesmus quadricauda* is unified based upon conventional results.

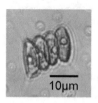

Tolypothrix sp.

In the thread-like *Tolypothrix* sp., which is related to blue-green algae, there is one trichome *per* sheath. The algal body is 5–10 mm in length, they float or cluster, and they become habitats for *Aeolosoma hemprichi* and *Philodina erythrophthalma* in the microcosm. Due to the difficulty in counting individuals, they are counted after crushing a colony in a homogenizer. The population constitutes mere centimeters in the microcosm.

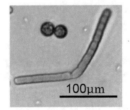

3.3.2 Micro-animal Consumers

Cyclidium glaucoma

Cyclidium glaucoma is a ciliate protozoan that feeds on bacteria. The sizes of the cells range from 15 to 25 μm in length, are ovoid in shape, and are 10 μm in width. The growth rate of this species is the fastest among all of the predators in the microcosm and decreases gently after peaking during the transition period on the 4th day after culturing begins (onset of culturing: 0–14 days after). It moves quickly, but it is necessary to carefully watch the ciliary movement even on the occasions in which it stops and to count it when it moves or stops suddenly. It may be treated as a rare species in the natural ecosystem based on previous evaluations (which have shown that there is an influence even if the species disappears) of a structural parameter for the addition of chemical substances and changes in environmental conditions.

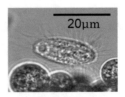

Lecane sp.

Lecane sp. is a rotifer that feeds on bacteria. Its body length ranges from 50 to 100 μm, and its body is oval in shape. It has a tail, and there are no corners. It feeds while swimming, but the movement is slow. "*Lepadella* sp." as described in the documentation of the microcosm is indicative of this species.

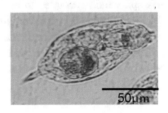

Philodina erythrophthalma

Philodina erythrophthalma is also a rotifer that feeds on bacteria that swim and crawl. Its body length is 300–500 µm; they generate a strong current using the ciliary ring at the anterior end, eat bacteria, and multiply. When population sizes increase parthenogenically (a female develops from a diploid summer egg), and biotope conditions worsen, males develop from the haploid egg, which are produced by a reduction in division that typically forms winter eggs (pause eggs). The walls of winter eggs are dryer and thicker than those of summer eggs when resistant to fertilization. As biotope conditions improve, a female then develops from a winter egg. The limbs adjust such that there are two of them at the posterior end, and there are four hocks.

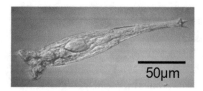

Aeolosoma hemprichi

Aeolosoma hemprichi is a detritivorous oligochaete (i.e., an aquatic earthworm). Its body length ranges from 500 to 1000 µm, and it appears in natural ecosystems, such as lakes and marshes, and in artificial ecosystems, such as the wastewater treatment systems of activated sludge and in biofilms. It exhibits a subcutaneous, crimson grease spot and has bunched bristles on each metamere for both of its two pairs of stomachs. This species eats bacteria and other microbes, such as protists and flocculated cells, multiplies by cell division, and excretes fecal pellets.

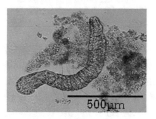

3.3.3   Bacterial Decomposers

Bacillus cereus

Bacillus cereus is a large, gram-positive aerobic bacterium in the *Bacillus* genus that has spores. It is abundant in the natural world, including in soils and polluted waters, and its growth is poor under acidic conditions. It causes food poisoning, and, as an indigenous bacterium, it is observed within the intestinal tract of 10% of healthy adults. Spores grow between 4 and 50 °C and germinate at 1–59 °C; heating it to 100 °C for 10 min can inactivate most spores, but they are able to endure heating to 100 °C for up to 30 min. This species is widely distributed in natural environments, and spores form in the soil.

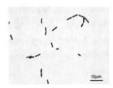

Pseudomonas putida

Pseudomonas putida is a gram-negative bacillus and is a soil microbe that is saprophytic. It is seen in the natural environment (i.e., in the soil and hydrosphere) when oxygen is present. The optimum growth temperature ranges from 25 to 30 °C, and some form colonies in plant rhizospheres. The surface of plant roots supplies nutrients to the bacteria, and *Pseudomonas putida* contributes to the growth of the plant by preventing pathogens from spreading in the roots. Use of this bacterium as a biological pesticide is expected, and strains of *Pseudomonas putida* are researched and developed for their ability to promote plant growth.

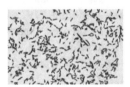

Acinetobacter sp.

Acinetobacter is a eubacterial genus of gram-negative bacilli. It thrives in wet environments, including soils, and is widely distributed across natural environments. It may also be present on the skin of healthy people, and it may be separated from the excrement of animals. It is aerobic and has a short, stick-like shape. It is fixed in place as it lacks flagella, it is oxidase negative and does not undergo fermentation with glucose, and it is relatively robust to drying. It is also typically harmless; however, there are also causal pathogens of opportunistic infectious diseases. Because it has a mechanism by which it is able to incorporate other DNA fragments into its own DNA, it may be considered a mutation-prone bacterium.

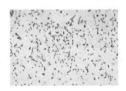

Coryneform Bacteria

The coryneform bacterium in the microcosm represents the eubacterial genus *Actinomyces*, with a gram-positive bacillus. The form is clavate, such that it is a bacillus with a straight or slightly curved tip, and one or both ends of the cell body often bulge (as is shown in the image on the right). It is relatively closely related to tubercle bacilli, and the cell wall has the characteristic of containing lipids called mycolic acids. Coryneform bacteria are broadly distributed across the natural world, and certain kinds are habitually present in the respiratory tract and mucous membranes of *Homo sapiens*.

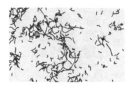

In addition to the described species, Table 3.2 shows that many species are known to coexist within gnotobiotic-type, flask-sized systems, such as the microcosm N-system. Of these, when the microcosm of the three-species system is excluded,

Table 3.2 Species composition of flask size gnotobiotic-type microcosm

	Producer	Consumer	Decomposer	Appendix
Odum	*Chlorella* sp.	*Paramecium bursaria*	*Pseudomonas* spp.	Ecological microcosms (Beyers and Odum)
	Schizothrix calcicola	*Lepadella* sp.	*Bacillus* spp.	
	Scenedesmus sp.	*Philodina* sp.	*Flavobacterium* spp.	
		Cypridopsis vidua		
Sugiura (Kurihara)	*Chlorella*	*Cyclidium*	More than 5 species	Papers
	Schizothrix	*Lepadella*		
		Philodina		
		Aeolosoma		
N system	*Chlorella* sp.	*Cyclidium glaucoma*	*Pseudomonas putida*	Stock culture made by Dr. Yasushi Kurihara transferred from Dr. Katsura Sugiura's Lab
	Scenedesmus quadricauda	*Lecane* sp.	*Bacillus cereus*	
	Tolypothrix sp.	*Philodina erythrophthalma*	*Acinetobacter* sp. coryneform bacteria	
		Aeolosoma hemprichi		

(continued)

Table 3.2 (continued)

	Producer	Consumer	Decomposer	Appendix
N type (Inamori)	*Chlorella* sp.	*Cyclidium glaucoma*	*Pseudomonas putida*	Mixed S type and K type by Dr. Yuhei Inamori's Lab
	Tolypothrix sp.	*Lepadella* sp.	*Bacillus cereus*	
		Philodina sp.	*Acinetobacter* sp. coryneform bacteria	
		Aeolosoma hemprichi		
S type (Shikano)	*Chlorella* sp.	*Cyclidium glaucoma*	*Pseudomonas putida*	Constructed by Dr. Shuichi Shikano at Dr. Yasushi Kurihara's Lab
		Lepadella sp.	*Bacillus cereus*	
		Aeolosoma hemprichi	*Acinetobacter* sp. coryneform bacteria	
K type (Shikano)	*Chlorella* sp.	*Lepadella* sp.	*Pseudomonas putida*	Constructed by Dr. Shuichi Shikano at Dr. Yasushi Kurihara's Lab
	Tolypothrix sp.	*Philodina* sp.	*Bacillus cereus*	
		Aeolosoma hemprichi	*Acinetobacter* sp. coryneform bacteria	
T type (Inamori&Tanaka)	*Chlorella vulgaris*	*Cyclidium glaucoma*	*Pseudomonas putida*	Constructed by Dr. Nobuyuki Tanaka at Dr. Yuhei Inamori's Lab
O type (Inamori&Murakami)	*Chlorella* sp.	*Cyclidium glaucoma*	*Pseudomonas putida*	Constructed by Dr. Kazuhito Murakami at Dr. Yuhei Inamori's Lab
	Oscillatoria agardhii	*Lepadella* sp.	*Bacillus cereus*	
		Philodina sp.	*Acinetobacter* sp. coryneform bacteria	
		Aeolosoma hemprichi		
Kawabata	*Euglena gracilis* Z	*Tetrahymena thermophila* B	*Escherichia coli* DH5 α	Same as Dr. Shoichi Fuma's Lab

the Kurihara-type and the Odum-type are confirmed, but their respective combinations of microbial compositions become similar. High plasticity in the microcosm N-system (Kurihara-type) is indicated by the stability of the system, which contains a food chain like in a natural ecosystem that is comprised of a producer, consumers, and a decomposer. Therefore, it is thought that this is appropriate as a model of the natural ecosystem when scaled down to flask size.

Literature Cited

Holt JG. Bergey's Manual of Determinative Bacteriology. 9th ed: Pennsylvania, Lippincott Williams & Wilkins; 1994.

Japan Sewage Works Association. Wastewater Examination Method -FY2012-; Tokyo, 2012 (in Japanese).

Shiga Scientific Teaching Materials Research Committee. Jananese Freshwater Plankton Illustrated Handbook. Tokyo, GodoShuppan, 150pp; (2005). (in Japanese).

Sudo R, Inamori Y. Diagnosis of Water Treatment from Microbiota. Tokyo. The Industrial Water Institute, 287pp; 1997. (in Japanese).

Tanaka M. An Illustrated Reference Book of Japanese Freshwater Plankton. Nagoya, The University of Nagoya Press, 584pp; 2002. (in Japanese).

Chapter 4
Preparation of the Microcosm N-System

Ken-ichi Shibata, Kunihiko Kakazu, Kakeru Ruike, and Kazuhito Murakami

Abstract In this chapter, the preparation of the microcosm test, including the apparatus, the dissolved oxygen (DO) meters (electrode DO meter and fluorometric DO meter), the calibration of the DO meter, and the oxygen dissolution rate are described. Additionally, the treatment of the microcosm N-system from successive subculturing is also explained.

4.1 Apparatus

- 300 mL Erlenmeyer flask
- Silicon plug (Matsuura Machinery, Catalog No. 6-343-16, Type No. C-40)
- Incubator (25 °C, L/D cycle = 12 hr/12 hr, illuminance = 2400 lux (photosynthesis photon flux density = 36 μmol/m^2/s))
- Dissolved oxygen (DO) meters
- Black rubber tube (only needed when light leaks from DO meter)
- Stirrer (unnecessary if the DO meter is able to perform measurements in standing water)
- Stirring rotor bar (unnecessary if the DO meter is able to perform measurements in standing water)

K. Shibata (✉)
Yokohama National University, Yokohama, Kanagawa, Japan
e-mail: shibata-kenichi-ym@ynu.jp

K. Kakazu
Foundation for Advancement of International Science, Tsukuba, Ibaraki, Japan
e-mail: kakazu@fais.or.jp

K. Ruike
University of Tsukuba/Foundation for Advancement of International Science, Tsukuba, Ibaraki, Japan
e-mail: ruike@fais.or.jp

K. Murakami (✉)
Chiba Institute of Technology, Narashino, Chiba, Japan
e-mail: kazuhito.murakami@p.chibakoudai.jp

© Springer Nature Singapore Pte Ltd. 2020
Y. Inamori (ed.), *Microcosm Manual for Environmental Impact Risk Assessment*,
https://doi.org/10.1007/978-981-13-6798-4_4

- Perforated cap for DO electrode (unnecessary if the DO meter is able to perform measurements in standing water)
- Vinyl tube (unnecessary if the DO meter is able to perform measurements in standing water)
- Circular glass cover (unnecessary if the DO meter is able to perform measurements in standing water)
- Insulating material (e.g., silicon sponge sheet) and blank paper or a white acrylic plate (unnecessary if the DO meter is able to perform measurements in standing water)

4.2   Dissolved Oxygen (DO) Meter Protocols

4.2.1   Electrode DO Meter

1. Wash the tip (gold cathode, silver anode) of the DO meter, a cap, an O-ring, and a rotor well in pure water (demineralized water), and ensure that they are dry before proceeding.
2. Remove any color change from the cathode by polishing its surface. Polish horizontally with a super fine-grit sandpaper (greater than a #1000) to clean the gold cathode. If the electrode is not smoothed, it will cause the membrane to tear during attachment.
3. When refuse collects in the ditch around the gold cathode, remove it using the tip of a toothpick. Additionally, this prevents touching of the cathode, as the triangular part of the electrode is very fragile.
4. Affix an electrode and a plastic bottle to the meter as shown in Fig. 4.1, and wind plastic tape around it so that it will not leak from around the adjoining part.
5. Leave undisturbed for 2–3 min until there remains 14% of the ammonium solution or 1–2 h until there remains 3% of the ammonium solution, with the

Fig. 4.1 Wash of electrode

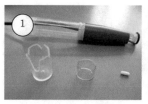

Fig. 4.2 Setup of DO electrode

tip of the DO meter completely submerged. If a terminal end has black dirt on it and does not become whitish, leave undisturbed for additional time.

6. Wash away the ammonium solution in pure water. If it does not wash well, dip it into pure water for a while.
7. Remove shavings by washing well with pure water, and then rinse it out three times in a potassium chloride solution (KCl).
8. Fill the tip with KCl and attach a membrane. Confirm that the film is not wrinkled and that air bubbles are not present inside of the membrane. When air bubbles are present, reattach the membrane. Remove it by pulling the edge of the membrane with an O-ring when a wrinkle is observed. Cut away any extra membrane afterward.
9. Place the glass cover and a stirring rotor bar into the cap, and attach it to the tip of the electrode. Prevent the silver electrode from touching the cap.

When using electrodes that require water flow to measure the DO (e.g., polarographic oxygen electrodes), use a perforated cap with a stirring rotor bar in the sensor region, and stir only near the sensor to measure DO. A stirring rotor bar can be moved smoothly by placing a circular glass cover at the bottom of the perforated cap and placing the rotor on it. A method of attaching this stirring system to the YSI-5750 is shown in Fig. 4.2. Insert the electrode (within which the stirring system has been emplaced) until it is in contact with the bottom of the 300 mL Erlenmeyer flask, and affix it using a rubber or silicon gasket or clip. Do not block the mouth of the flask. Attach an insulating material and blank paper to the stirrer so that the heat of the stirrer will not reach the flask. A white acrylic board (which has been processed to a thickness of 5 mm and has a gap of 5 mm formed below it) may be substituted for the insulating material and blank paper. Carry the flask equipped with an electrode on this board, stir it, and then measure the DO.

4.2.2 Fluorometric DO Meter

Because light leaks out from the side of the electrode, a fluorescence-type DO meter, such as the ProODO (YSI, Japan), is used to interrupt light in a black rubber tube (Fig. 4.3). Affix this electrode using a rubber or silicon gasket or a clip from the

Fig. 4.3 Interruption of
light in a black rubber tube

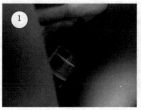

bottom of the 300 mL Erlenmeyer flask to a depth of ~1 cm. Do not block the mouth of the flask. Stirring of the flask is not necessary, as spatial inhomogeneity is considered an ecological characteristic.

Additionally, it is appropriate to use the fluorescence-type (optical) DO meter in the microcosm test due to its ease of maintenance and stable performance.

4.2.3   Calibration of the DO Meter

Pour 210 mL of distilled water or Milli-Q water into a 300 mL Erlenmeyer flask, attach a DO electrode in the manner prescribed in Sects. 4.2.1 and 4.2.2, and measure DO continuously overnight every 30 min. Review the DO concentration to confirm that it is stable, and then calibrate the concentration at that time as 100%. Additionally, finish calibrating atmospheric pressure before the calibration of DO when there is an automatic correction function based on the atmospheric pressure.

4.2.4   Oxygen Dissolution Rate

The rate (δD) at which oxygen moves between the atmosphere and the water is influenced by the state of the electrode, the incubator, the stirrer, the opening of the mouth of the flask, and other variables. Therefore, every experimental replicate (i.e., the flask, electrode, stirrer, etc.) should be tested, and the oxygen dissolution rate should be determined individually or in bulk by calculating the average. If possible, it is desirable to measure δD twice, immediately before and immediately after an experiment, and to obtain the mean DO concentration; δD should be measured according to the protocol described below.

4.2.4.1   Experiment

Pour 210 mL of distilled or Milli-Q water into a 300 mL Erlenmeyer flask, and aerate with oxygen or nitrogen. Measure the DO concentration of the liquid phase so that it is higher than the equilibrium level in the atmosphere (in the case of oxygen) or so

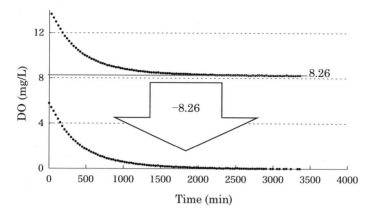

Fig. 4.4 Reduction of 8.26 from measured value

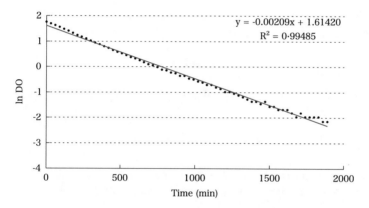

Fig. 4.5 Linear approximation by logarithm with deduction of 8.26

that it is lower (in the case of nitrogen). Leave the flask undisturbed for ~60 min without closing the cap. Then, replace the oxygen or nitrogen of the vapor phase in the flask with air, and measure the DO for 24–48 h, as described in Sect. 4.2.3.

The method of calculating the δDO with data when aerated with oxygen is shown in Fig. 4.4 as an example. The black point in the upper section of Fig. 4.4 represents the DO data when aerated with oxygen. Dissolved oxygen decreases exponentially and reaches equilibrium at 8.26 mg/L. When 8.26 mg/L is subtracted from the original DO data, it results in the graph shown in the bottom panel of Fig. 4.4.

The linear approximation of the natural logarithm ($\ln x = \log_e x$) for the data (>0) subtracted from 8.26 mg/L is shown in Fig. 4.5. However, large errors arise as the line approaches the equilibrium concentration, and, after approximately 700 min, it cannot be approximated well because of the amount of noise.

The slope of the approximately straight line shows the constant rate of oxygen transfer. Assuming that the constant rate of oxygen transfer is a and the equilibrium

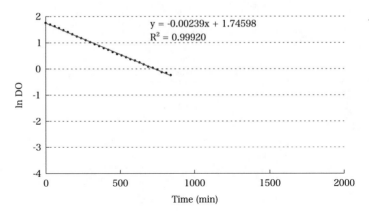

Fig. 4.6 Linear approximation with removal of measurement error

concentration of DO is b, the oxygen transfer rate (δDO) at a certain DO concentration x can be calculated according to the following formula:

$$\delta DO = a(x - b).$$

Since $b = 8.26$ mg/L (Fig. 4.4) and $a = -0.00239$ (Fig. 4.6), δDO (mg/L/min) in this example is

$$\delta DO = -0.00239\,(x - 8.26).$$

This oxygen transfer rate (δDO) is used in Sect. 6.2, "Calculation of Respiration and Production Quantities in the Microcosm System," of this book.

4.3 Microcosm N-System

The maintenance culture of the microcosm N-system must remain aseptic. The culture used for experimentation adhered to a culture maintenance method. The inoculum dose of the N-system seed is 10 mL. A TP (Taub + peptone) medium (shown in Table 3.1) is employed for cultivation. After adjusting the TP medium, dispense it into a 300 mL Erlenmeyer flask by 200 mL volume, cap it with a silicon plug, and then autoclave it at 121 °C for 15 min to sterilize it. After allowing it to cool to room temperature, add 10 mL of the N-system, which was cultured for 2–8 weeks, as a seed, and culture it in an incubator with an L/D cycle = 12 hr/12 hr and an illuminance = 2400 lux (photosynthesis photon flux density = 36 μmol/m²/ s), at a temperature of 25 °C. Stir it lightly once a day until the culture begins 1–3 days later, then leave it at rest, and culture it afterward. Subculturing is performed once every 2 months. Use a National (i.e., Panasonic) FL20SS-W/18 as

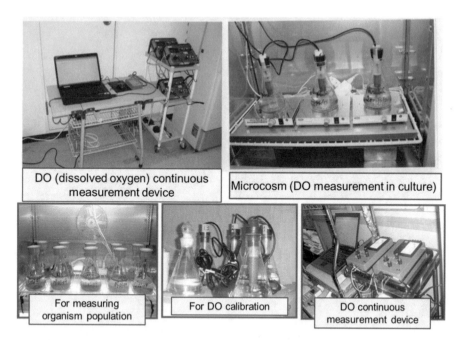

Fig. 4.7 Microcosm N-system apparatus

a source of light, replace all fluorescent bulbs every 3 months or measure the illuminance or photosynthesis photon flux density every month, and replace them as needed. Various devices associated with culturing of the microcosm are shown in Fig. 4.7. The DO electrode that was attached to the microcosm was connected to a DO meter outside of the incubator, and the wave patterns of the DO were continuously monitored using a data logger on a personal computer at the same time as they were recorded. When an abnormality is observed within a wave pattern, immediately reset the DO sensor. Preparation of the N-system used for biomass and chemical analyses and the one used for measuring DO should be performed separately. Additionally, both microcosms should be cultured equally until an electrode for DO measurement is attached.

Literature Cited

Kurihara Y. Studies of Succession in a Microcosm, Science Report of Tohoku University, 4 th Ser. (Biology), No. 37; 1978a. p. 151–160.

Kurihara Y. Studies of the Interaction in a Microcosm, Science Report of Tohoku University, 4 th Ser. (Biology), No. 37; 1978b. p. 161–177.

Taub FB. Measurement of Pollution in Standardized Aquatic Microcosms. In: White HH, editor. Concept in Marine Pollution Measurements. College Park: University of Maryland; 1984. p. 139–58.

Chapter 5
Procedure for Using the Microcosm N-System

Ken-ichi Shibata, Kunihiko Kakazu, Kakeru Ruike, and Kazuhito Murakami

Abstract In this chapter, the measured biomass, amount of production and consumption, an endpoint, and a reference point for ecosystem risk assessment for the microcosm are presented.

5.1 Work Flow for the Microcosm N-System

The work flow for the impact assessment using the experimental microcosm is as follows:

1. Preparations of the microcosm, test materials, metering equipment, and culture apparatus
2. Initiation of the microcosm culture
3. Initiation of DO measurements (0 or 15 days after initiation of microcosm cultivation)
4. Addition of the analyte
5. End of cultivation and measurement (14 days after the analyte was added to the microcosm)

K. Shibata
Yokohama National University, Yokohama, Kanagawa, Japan
e-mail: shibata-kenichi-ym@ynu.jp

K. Kakazu
Foundation for Advancement of International Science, Tsukuba, Ibaraki, Japan
e-mail: kakazu@fais.or.jp

K. Ruike
University of Tsukuba/Foundation for Advancement of International Science, Tsukuba, Ibaraki, Japan
e-mail: ruike@fais.or.jp

K. Murakami (✉)
Chiba Institute of Technology, Narashino, Chiba, Japan
e-mail: kazuhito.murakami@p.chibakoudai.jp

© Springer Nature Singapore Pte Ltd. 2020
Y. Inamori (ed.), *Microcosm Manual for Environmental Impact Risk Assessment*,
https://doi.org/10.1007/978-981-13-6798-4_5

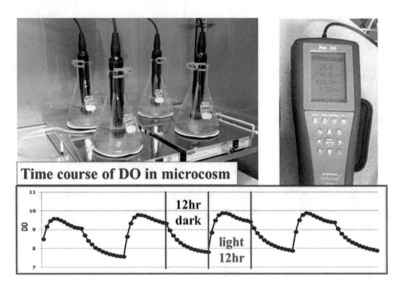

Fig. 5.1 As shown in Figures 5.1 and 5.2, install the DO electrode on the 14th day (336 hours) after the microcosm cultivation, and then start the DO measurement

6. Cleanup and organization
7. Statistical analysis of the resultant data (significance determined by a branching analysis of variance)
8. Determination of whether or not there was any influence on the microcosm using the impact assessment criteria

One full test can be completed in 30 days. However, because pre-cultivation of the microcosm is necessary, culturing should be repeated several times before the final assessment, and it should be confirmed that the control system (in which no analyte was added) formed a stable system (Fig. 5.1). Microcosm N-system with DO electrode and DO fluctuating pattern in microcosm were shown in Fig. 5.1.

5.2   Setup of a DO Meter in the Microcosm N-System

Install the DO electrode at the 14th culture time (336 h) and begin the measurements. Setup of the electrode follows the methods described in Sects. 4.2.1 and 4.2.2. Care should be taken when installing the electrode to prevent disturbance of the distribution of organisms in the microcosm. After setup, it is not necessary to operate it aseptically when installing an electrode. However, the electrode should be sterilized with alcohol to avoid trouble with the electrode, which may be installed after confirming by smell that the alcohol has dried (Fig. 5.2).

Fig. 5.2 Fixation of electrode

5.3   Addition of Substances to the Microcosm N-System

Addition of test materials to the microcosm is conducted on day 0 (i.e., when microcosm cultivation was initiated) or on the 16th day (i.e., when the microcosm reached a stationary phase). During the addition of test materials on day 0, the effect of test materials on neither the microcosm nor the microbiota was evaluated (i.e., it was not determined if a stable ecosystem could be established or not). Similarly, during the addition of test materials on the 16th day, the effect of the test materials on neither the microcosm nor the microbiota was evaluated (i.e., it was not determined whether the species diversity was maintained or whether a stable ecosystem could be recovered). For both addition methods, the microbial population (structural parameter) and the P/R ratio (functional parameter) should be measured to evaluate the effects of the test materials.

5.3.1   Replication and Control Groups

Preliminary examinations conducted 1–3 times are sufficient. Repeat the final experiment 5 times; in consideration of the statistics of the analyzed data, the final experiment should be performed *at least* three times. Add a solvent if the same amount as that used in the test group is used in the control group. When a material (e.g., organic solvents, such as acetone) other than pure water (demineralized water) is used as a solvent, both the pure water and solvent should be added to the control group.

5.3.2 Preparation of Test Substances

Allow the test material to dissolve in distilled or Milli-Q water. When it is difficult to dissolve, scatter it via sonication. If no other method is used, dispersants such as inorganic acids and bases, HCl, NaOH, or acetone, Cremophor RH40 and 0.01% methylcellulose aqueous solution, HCO-50, HCO-60, and others (<100 µL/L) are used; however, this should be minimized. Regulate the concentration of the analyte after addition to the microcosm; this then becomes the designated test concentration in the microcosm.

5.3.3 Addition of Test Substances (During the Stationary Phase)

The test material is added during culturing for 360 h on the 16th day. Insert glass tubes, such as Pasteur pipettes, into the N-system to measure DO from the gap between an electrode and the flask. Carefully drain the solution of the test material into the Pasteur pipette and add it. Drain a solvent equal to triple the volume of the solution into a Pasteur pipette, and then add all of the test materials to the N-system. Avoid disturbing the distribution of organisms in the microcosm; there is no need to stir. If this method is difficult, the test material can be added by removing the electrode once. Once again, avoid disturbing the distribution of the organisms in the microcosm. The DO measurements are not influenced by the exclusion or addition of the electrode or test materials.

Carefully add the test materials using a pipette to the open mouth of the microcosm N-system for analyses, and avoid disturbing the distribution of the organisms within it. There is no need to stir the flask. The volume of the solvent and that of the test material are even for DO measurements. Prepare the control group to which water was added. If a solvent other than water is used, the influence of the solvent must be confirmed by preparing a control group to which solvent is added separately from the control group to which water was added.

5.3.4   Addition of Test Substances (During the Growth Phase)

Add test materials to an empty and sterile 300 mL Erlenmeyer flask. When an organic solvent is used, absorb it and remove it. Add 200 mL of fresh TP_{100} nutrient medium into this flask, and inoculate 10 mL of the microcosm N-system from the plateau stage, and begin cultivation. Install a DO electrode and begin measurement from a cultivation initiation time as described in Sect. 5.3.3. Additionally, for ecosystem impact assessments of various chemical substances, the method described in Sect. 5.3.3 is generally used.

Chapter 6
Estimation Using the Microcosm N-System

Kunihiko Kakazu, Kakeru Ruike, Ken-ichi Shibata, and Kazuhito Murakami

Abstract In this chapter, the methods for measuring biomass and calculating the amount of production and consumption and an endpoint for a standard ecosystem impact assessment are described.

6.1 Biomass Measurement in the Microcosm N-System

Biomass should be measured by the method described herein 14 days (336 h) after the addition of the test materials, and the quantity of the test materials should be measured at the same time.

The constituent species of this microcosm are as follows:

- *Aeolosoma hemprichi* (oligochaete predator/detritus feeder)
- *Philodina erythrophthalma* (rotifer predator/bacteria feeder)
- *Lecane* sp. (rotifer predator/bacterivore)
- *Cyclidium glaucoma* (ciliate predator/bacterivore)
- *Chlorella* sp. (chlorophycean producer)
- *Scenedesmus quadricauda* (chlorophycean producer)
- *Tolypothrix* sp. (cyanophycean producer)

K. Kakazu
Foundation for Advancement of International Science, Tsukuba, Ibaraki, Japan
e-mail: kakazu@fais.or.jp

K. Ruike
University of Tsukuba/Foundation for Advancement of International Science, Tsukuba, Ibaraki, Japan
e-mail: ruike@fais.or.jp

K. Shibata
Yokohama National University, Yokohama, Kanagawa, Japan
e-mail: shibata-kenichi-ym@ynu.jp

K. Murakami (✉)
Chiba Institute of Technology, Narashino, Chiba, Japan
e-mail: kazuhito.murakami@p.chibakoudai.jp

© Springer Nature Singapore Pte Ltd. 2020   45
Y. Inamori (ed.), *Microcosm Manual for Environmental Impact Risk Assessment*,
https://doi.org/10.1007/978-981-13-6798-4_6

– Bacteria (at least four species, dominated by *Pseudomonas putida*, *Bacillus cereus*, *Acinetobacter* sp., and coryneform bacteria)

Aeolosoma hemprichi, *Philodina erythrophthalma*, *Lecane* sp., *Cyclidium glaucoma*, *Chlorella* sp., *Scenedesmus quadricauda*, and *Tolypothrix* sp. must be counted individually. However, the algal biomass can be substituted for chlorophyll *a* and *b*. In the case of substitution of the algal biomass for chlorophyll *a* and *b*, it is desirable to quantitatively measure phycocyanin, phycoerythrin, and allophycocyanin.

6.1.1 Aeolosoma hemprichi *and* Philodina erythrophthalma

The number of *Aeolosoma hemprichi* and *Philodina erythrophthalma* should be counted under an optical microscope using a glass slide with a grid as a borderline. Independent measurements should be performed more than three times and with a total of more than ten individuals in 5 mL or less. The unit of measure is N/mL.

6.1.2 Lecane *sp. and* Cyclidium glaucoma

A hemocytometer should be used. to count the number of *Lecane* sp. and *Cyclidium glaucoma* with a phase contrast microscope. Independent measurements should be performed more than three times and with more than 20 individuals in total in 300 μL or less. The unit of measure is N/mL.

6.1.3 Chlorella *sp. and* Scenedesmus quadricauda

As *Chlorella* sp. and *Scenedesmus quadricauda* exist. as both single cells and colonies, a hemocytometer should be used to count them with a phase contrast microscope after disaggregating the colonies by sonication (20 s). Independent measurements should be performed more than three times and measured until the more abundant alga reaches 400 cells in total. The unit of measure is cells/mL.

6.1.4 Tolypothrix *sp.*

Because identifying species. of *Conferva* (filamentous algae) is difficult and the measurement accuracy decreases slightly when sonicated, it is best not to sonicate it before measurements are taken. Many *Tolypothrix* sp. become twisted, further complicating their measurement, and can only be assayed after processing for 10 s with a weak output (10 w). The use (or lack thereof) of sonication and the output and time should be uniform across every test. The length of *Tolypothrix* sp. in the sample, which is sonicated as needed, is measured through a digital microscope screen using a hemocytometer. If this is impossible, a micrometer should be used, or the length should be measured based on the approximate length of the slide grid. The unit of measure is cm/mL.

6.1.5   Chlorophyll a *and* b

There are many ways of extracting and assaying chlorophyll *a* and *b*, but an extraction method was introduced by Burnison (1980) that uses dimethyl sulfoxide (DMSO), which is generally efficient and easy to handle. Approximately 1–2 mL of filtered culture medium and a glass fiber filter (either GF/F from Whatman, UK, or a GF-75 from Advantec, Japan) 25 mm in diameter are used. Filter paper should be lightly rounded and placed in a centrifuge tube of 1.5–2.0 mL. The DMSO is then added and sonicated three times for 30 s. Once preserved, it may be measured later; it may either be preserved at -20 °C before the addition DMSO and sonication or it may be completely extracted and preserved at 4 °C. However, the preservation method for all samples must be uniform. If the sample has been refrigerated, it should be brought back up to room temperature and left to rest at 65 °C for 10 min. Centrifugal separation (i.e., at 12,000 rpm for 10 min) is then performed before the removal of the filter and cell debris. The spectrum of the supernatant should be measured with a spectrophotometer, and the concentration of chlorophyll *a* and *b* is then calculated using the following equation (Wellburn 1994):

$$A_\lambda = (\text{absorbance in wavelength } \lambda \text{ nm}) - (\text{absorbance in wavelength } 750 \text{ nm}) :$$
$$\text{Chl}\, a = 12.19 A_{665} - 3.45 A_{649}$$
$$\text{Chl}\, b = 21.99 A_{649} - 5.32 A_{665}$$

where A is the absorbance in nanometers and *Chl* is chlorophyll. Additionally, it can be assumed that the phaeophytization quotient (PQ level $= A_{435}/A_{415}$) is an index of the stress.

6.1.6   *Phycocyanin, Phycoerythrin, and Allophycocyanin*

A simple method of extracting and assaying C-phycocyanin (C-PC), C-phycoerythrin (C-PE), and allophycocyanin (APC) was developed by Patel et al. (2005). Approximately 5–10 mL of culture medium must be centrifuged, and the supernatant should be filtered by a 25 mm diameter glass fiber filter, either GF/F

Table 6.1 PY medium

Polypepton	5 g
NaCl	4 g
Yeast extract	3 g
Agar	20 g
Distilled water	1 L

(Whatman, UK) or GF-75 (Advantec, Japan). The filter paper should be rounded lightly, returned to the centrifuge tube, and combined with the precipitate. A potassium phosphate buffer (0.1 M, pH 7.0) is then added before being sonicated three times for 30 s. The prepared culture should be frozen at -20 °C and then melted to room temperature. In the case in which the culture is preserved once and measured later, it should either be preserved at 4 °C after complete extraction or at -20 °C with the precipitate and filter combined. However, the preservation method must be uniform for all samples. After centrifugal separation (i.e., at 12,000 r.p.m. for 10 min), the filter and cell debris should be removed, and the spectrum of the supernatant should be measured with a spectrophotometer. The concentrations of C-PC, C-PE, and APC may then be calculated from the following equation (Bennet and Bogorad 1973) (Table 6.1):

$$A_\lambda = \text{ absorbance in wavelength } \lambda \text{ nm}$$
$$\text{C-PC} = (A_{615} - 0.474A_{652})/5.34$$
$$\text{APC} = (A_{652} - 0.208A_{615})/5.09$$
$$\text{C-PE} = (A_{562} - 2.41 \text{ C-PC} - 0.849 \text{ APC})/9.62$$

6.1.7 Bacteria

After taking 0.5–1 mL of samples and performing sonication (30 s), the abundance of living bacteria should be measured, according to the colony calculation method using the peptone-yeast (PY) extract nutrient medium, by adjusting the number of bacteria through serial dilution. The cultivation period for colony forming unit (CFU) counting is set to 5 days at 30 °C. A 30–300 CFU/plate (ideal: 100 CFU/plate) plate should be chosen, and 2–3 pieces should be measured. The unit of measure is CFU/mL. In the case of using an antibiotic as a test material, it is desirable to assay the drug-resistant bacteria for the antibiotic concerned. In the case of a detailed analysis, the biomass calculation described here is necessary; however, in typical evaluations, measurements of the P/R ratio are generally used.

6.2   Calculation of Respiration and Production Volumes in the Microcosm N-System

6.2.1   Outline

The amount of respiration (R) and total amount of production (P) are calculated from the DO concentration between 312 and 336 h after the addition of a test material to the microcosm N-system. Generally, the amount of respiration (R), net amount of production (Pn), and total amount of production (P) can be calculated from an increment (ΔDO) of the DO concentration by the following expressions:

$$\Delta DO_{D/L} = \text{total increment of the DO in the L/D photoperiod} \qquad (6.1)$$

$$R = -2 \times \Delta DO_D \qquad (6.2)$$

$$\text{Pn} = \Delta DO_L \qquad (6.3)$$

$$P = \text{Pn} + 1/2R \qquad (6.4)$$

There is an exchange of oxygen with the atmosphere regardless of biological activity, as calculated by the dissolution rate of oxygen (δDO), described in Sect. 4.2.4. Therefore, it is necessary to revise expressions (6.2) and (6.3) as (6.6) and (6.7), respectively.

$$\delta DO_{D/L} = \text{total solubility from the atmosphere to culture medium for an } L/D \text{ photoperiod}$$
$$(6.5)$$

$$R = -2 \times (\Delta DO_D - \delta DO_D) \qquad (6.6)$$

$$\text{Pn} = \Delta DO_L - \delta DO_L \qquad (6.7)$$

$$P = \text{Pn} + 1/2\text{R}$$

$$P = \text{Pn} + 1/2R. \qquad (6.4)$$

6.2.2   Calculation

The methods for calculating ΔDO_D and ΔDO_L and δDO_D and δDO_L in MS Excel are shown in Figs. 6.1 and 6.2, respectively. Figure 6.1 shows a numerical formula to input into each cell, and Fig. 6.2 shows the calculation results. Additionally, although the δDO, which was calculated in Sect. 4.2.4, was -0.00239 ($x - 8.26$) mg/L/min, these data were measured in 1-hour increments and became -0.00239 ($x - 8.26$) $\times$ 60 $= -0.1434$ ($x - 8.26$) mg/L/h. From Fig. 6.3, it can be seen that

	Time	DO	ΔDO	Total of ΔDO	Average of DO	δDO	Total of δDO
	1	8.78					
	2	8.56	=C3-C2		=AVERAGE(C2:C3)	=-0.1434 * (F3-8.26)	
	3	8.32	=C4-C3		=AVERAGE(C3:C4)	=-0.1434 * (F4-8.26)	
	4	8.16	=C5-C4		=AVERAGE(C4:C5)	=-0.1434 * (F5-8.26)	
	5	8	=C6-C5		=AVERAGE(C5:C6)	=-0.1434 * (F6-8.26)	
Dark cycle	6	7.86	=C7-C6		=AVERAGE(C6:C7)	=-0.1434 * (F7-8.26)	
	7	7.78	=C8-C7		=AVERAGE(C7:C8)	=-0.1434 * (F8-8.26)	
	8	7.7	=C9-C8		=AVERAGE(C8:C9)	=-0.1434 * (F9-8.26)	
	9	7.64	=C10-C9		=AVERAGE(C9:C10)	=-0.1434 * (F10-8.26)	
	10	7.62	=C11-C10		=AVERAGE(C10:C11)	=-0.1434 * (F11-8.26)	
	11	7.58	=C12-C11		=AVERAGE(C11:C12)	=-0.1434 * (F12-8.26)	
	12	7.56	=C13-C12	ΔDO_D	=AVERAGE(C12:C13)	=-0.1434 * (F13-8.26)	δDO_D
	13	7.52	=C14-C13	=SUM(D3:D14)	=AVERAGE(C13:C14)	=-0.1434 * (F14-8.26)	=SUM(G3:G14)
	14	8.7	=C15-C14		=AVERAGE(C14:C15)	=-0.1434 * (F15-8.26)	
	15	9.6	=C16-C15		=AVERAGE(C15:C16)	=-0.1434 * (F16-8.26)	
	16	10.03	=C17-C16		=AVERAGE(C16:C17)	=-0.1434 * (F17-8.26)	
	17	10.05	=C18-C17		=AVERAGE(C17:C18)	=-0.1434 * (F18-8.26)	
Light cycle	18	9.86	=C19-C18		=AVERAGE(C18:C19)	=-0.1434 * (F19-8.26)	
	19	9.66	=C20-C19		=AVERAGE(C19:C20)	=-0.1434 * (F20-8.26)	
	20	9.48	=C21-C20		=AVERAGE(C20:C21)	=-0.1434 * (F21-8.26)	
	21	9.32	=C22-C21		=AVERAGE(C21:C22)	=-0.1434 * (F22-8.26)	
	22	9.18	=C23-C22		=AVERAGE(C22:C23)	=-0.1434 * (F23-8.26)	
	23	9.06	=C24-C23		=AVERAGE(C23:C24)	=-0.1434 * (F24-8.26)	
	24	8.94	=C25-C24	ΔDO_L	=AVERAGE(C24:C25)	=-0.1434 * (F25-8.26)	δDO_L
Dark cycle	25	8.86	=C26-C25	=SUM(D15:D26)	=AVERAGE(C25:C26)	=-0.1434 * (F26-8.26)	=SUM(G15:G25)

Fig. 6.1 Enter data to MS Excel sheet

because $\Delta DO_D = -1.26$ and $\delta DO_D = 0.60$, these values may be substituted for expressions (6.6), (6.7), and (6.4), and R, Pn, and P can be calculated as follows:

$$R = -2 \times (\Delta DO_D - \delta DO_D) = -2 \times (-1.26 - 0.60) = 3.72 \text{ mg/L/day}$$

$$Pn = \Delta DO_L - \delta DO_L = 1.34 - (-1.86) = 3.2 \text{ mg/L/day}$$

$$P = Pn + (1/2)R = 3.2 + 3.72 \times (1/2) = 5.06 \text{ mg/L/day}$$

6.3 Endpoint of the Microcosm N-System

It may be said that the endpoint of the toxicity test in the environmental impact assessment is an index for expressing the negative influence that a chemical substance generally has on an organism. The endpoints of this assessment using the microcosm N-system are production (P) and respiration (R), which allow for the influence of chemical substances (i.e., test materials) to be easily and appropriately determined. The biomass or chlorophyll a or b are measured secondarily and not as endpoints. Namely, a functional parameter (i.e., P, R, or P/R ratio) is defined as the

	Time	DO	Δ DO	Total of Δ DO	Average of DO	δ DO	Total of δ DO
	1	8.78					
	2	8.56	-0.22		8.67	-0.06	
	3	8.32	-0.24		8.44	-0.03	
	4	8.16	-0.16		8.24	0.00	
	5	8.00	-0.16		8.08	0.03	
Dark cycle	6	7.86	-0.14		7.93	0.05	
	7	7.78	-0.08		7.82	0.06	
	8	7.70	-0.08		7.74	0.07	
	9	7.64	-0.06		7.67	0.08	
	10	7.62	-0.02		7.63	0.09	
	11	7.58	-0.04		7.60	0.09	
	12	7.56	-0.02	Δ DO$_D$	7.57	0.10	δ DO$_D$
	13	7.52	-0.04	-1.26	7.54	0.10	0.60
	14	8.70	1.18		8.11	0.02	
	15	9.60	0.90		9.15	-0.13	
	16	10.03	0.43		9.82	-0.22	
	17	10.05	0.02		10.04	-0.26	
Light cycle	18	9.86	-0.19		9.96	-0.24	
	19	9.66	-0.20		9.76	-0.22	
	20	9.48	-0.18		9.57	-0.19	
	21	9.32	-0.16		9.40	-0.16	
	22	9.18	-0.14		9.25	-0.14	
	23	9.06	-0.12		9.12	-0.12	
	24	8.94	-0.12	Δ DO$_L$	9.00	-0.11	δ DO$_L$
Dark cycle	25	8.86	-0.08	1.34	8.90	-0.09	-1.86

Fig. 6.2 Calculation result by MS Excel (results of Fig. 3.2)

main endpoint, and a structural parameter (population size/abundance) is a secondary endpoint. Because the microcosm N-system is used in this study as a model of ecosystem function, it maintains dynamic equilibrium while transitioning (succession) among constituent species, as it always produces a natural ecosystem again. This is thought to be because it is proper to adopt the DO and P/R ratio, which is a functional parameter, for evaluation in the microcosm test. Additionally, because it is important to analyze and evaluate the assessment period from both the P/R ratio (functional parameter) and the population size/abundance (structural parameter), (1) it takes 2 weeks (14 days) for the microcosm to reach a steady state from the beginning of cultivation, and (2) once the constituent organisms of the microcosm reach a steady state, the system is maintained, and the population is held at a constant population from thereon. In other words, the recovery of the system cannot be expected if a stable system is not established by 2 weeks after the start of cultivation. It is proper that a period of 2 weeks after the addition of test materials (i.e., the 30th day if the materials were added on the 16th day after the start of cultivation) is used as the end of cultivation and that changes are evaluated during this period.

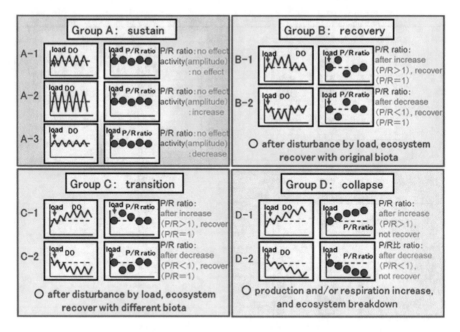

Fig. 6.3 Classification of P/R ratio succession pattern and determination of environmental impact

6.3.1 Statistical Processing (Branching-Type ANOVA)

The branching-type analysis of variance (ANOVA) used in the statistical processing of this microcosm assessment is a technique similar to a two-way parameter model*. In other words, the amount of change in the measurements over time is generally expressed as a fixed number (as the amount of change according to a certain rule). However, for analyzing the effects of a chemical substance when each test individual is randomly selected from a population, individual differences are considered to be probabilistic (i.e., differences in the resultant value are considered to be errors, even if all experimental conditions are matched). When there is a difference in the error produced by a certain element (individual difference) when the data for the experimental system is observed to have the same individuals under the same conditions, the difference is uncertain and is viewed as being subject to chance (i.e., error) (Japanese Red Cross Toyota Nursing University Extension 2005; Ono 2003; Kakidume 2008; Kataya and Matsufuji 2003; Ichihara 1990). Based on this notion, it was decided that a statistical analysis of P and R would be conducted using the branching-type ANOVA of the difference between the flask-sized experiments of each microcosm. For an assessment test, it was assumed that the difference would vary stochastically (i.e., as a random variable), unlike in the two-way fixed effect model of the microcosm test. This means that a control group and chemical substance addition group were randomly chosen from the population in which the microcosm was cultured, and the difference between each flask was considered to

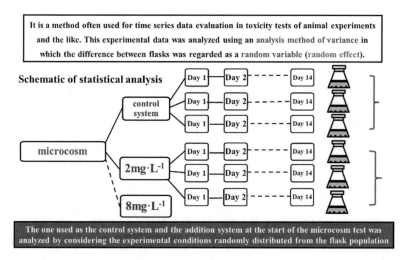

It is a method often used for time series data evaluation in toxicity tests of animal experiments and the like. This experimental data was analyzed using an analysis method of variance in which the difference between flasks was regarded as a random variable (random effect).

The one used as the control system and the addition system at the start of the microcosm test was analyzed by considering the experimental conditions randomly distributed from the flask population

Fig. 6.4 Outline of branching-type ANOVA in microcosm test

be due to error. Therefore, differences in production between flasks were considered to be caused by accident (i.e., error) and not by differences between the characteristics of the flasks. The outline of the branching-type ANOVA technique is shown in Fig. 6.4.

*A two-way parameter model is a method that assumes that the amount of change with respect to individual differences and the changes in response over time may be expressed as constants and that the population mean of each measured value can be expressed as the sum of these constants.

The microcosm is cultured under the same conditions allotted to either a control system or an addition system. While identical microcosms should have originally been generated, some differences are observed when they are actually measured because there are many coexisting species within the original system. These differences are not based upon any fundamental characteristics, and it is thought that they occur because of various accidental factors. Namely, such differences are assumed to be related to various accidental products and probabilistic errors, and the experimental microcosms are randomly assigned to either a control system or an addition system. From these randomly assigned microcosms, daily fluctuations are measured continuously.

In contrast with the individual differences among the microcosms described above, it is thought that some of the daily fluctuations in the microcosm data are in accordance with principles underlying the characteristics of the microcosms. In other words, when experimental microcosms from the original population are randomly assigned to the control system and the addition system, it is assumed that the differences between the flasks will fluctuate non-stochastically (i.e., nonrandomly). The branching-type ANOVA is the analytical method for measuring 14-day data that assumes the experimental conditions shown in Fig. 6.4. Additionally, this analysis is called a "branching-type" ANOVA (of the population from the above group of microcosms) because it branches, as shown diagrammatically in Fig. 6.4.

The model formula (1) of the branching-type ANOVA is shown below:

$$y_{ijk} = \mu + \alpha_i + b_{j(i)} + \gamma_k + u_{ijk}$$

Here, μ is the grand mean (i.e., the average of all data), αi is the main effect (i.e., the difference between the control system and the addition system), $b_{j(i)}$ is the difference between the flask and the assumption of randomness (i.e., stochasticity), γ_k is the main effect of daily fluctuations after the addition of a chemical substance, and u_{ijk} represents error. In other words, the grand mean of μ is the average of all of the data, such that:

(1) i represents the number of groups and is equal to 1 in the control system and 2 in the addition system.
(2) j represents the difference between the flask and shows the number of repeated experiment times, and it becomes 1 in the first experiment and 2 in the second experiment.
(3) k represents the days in progress after the addition of a chemical substance (i.e., test material); it becomes 1 for the first day after addition and 2 for the second day.
(4) It is assumed that certain measurement data, y_{ijk}, is the summed value of μ, α_i, $b_{j(i)}$, γ_k, and u_{ijk} and measurement data, y_{ijk}, for the third experiment, on the fourth day after the addition of a chemical substance, is expressed as $y_{234} = \mu + \alpha 2 + b_{3(2)} + \gamma_4 + u_{234}$.
(5) According to this, the measurement data y_{234} are judged as to whether (1) the test material addition (α_2) has an influence, (2) there is a difference between the flasks ($b_{3(2)}$), and (3) the changes over time (γ_4) are due to error (u_{234}).
(6) A branching-type ANOVA for higher addition concentrations is performed according to a closed test procedure, and the highest concentration with no recognized effect on the microcosm N-system is determined as the microcosm no-effect concentration (m-NOEC). The flow chart of branching-type ANOVA calculation is shown in Fig. 6.5.

The instructions for the practical calculation procedure are shown below.

− Procedure (1): From all of the data, y_{ijk}, calculate each mean. The following means (1–5) are the averages of each element:

   (1) Mean of each microcosm $\bar{y}_{ij.}; i = 1,2,\cdots, a; j = 1,2,\cdots, b$
   (2) Mean of each population $\bar{y}_{i..}; i = 1,2,\cdots, a$
   (3) Mean of each population in each time $\bar{y}_{i.k}; i = 1,2,\cdots, a; k = 1,2,\cdots, c$
   (4) Mean of each time $\bar{y}_{..k}; k = 1,2,\cdots, c$
   (5) Grand mean $\bar{y}_{...}$
        (population, a = 2, experiment, b = 3 or more, and time, c = 14, in this microcosm test method)

− Procedure (2): Calculate each sum of squares from the next expressions (1–8):

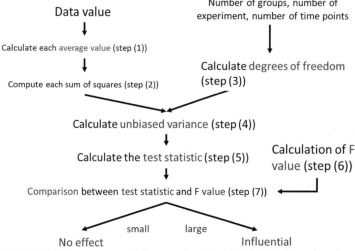

Fig. 6.5 Flow chart of branching-type ANOVA calculation

(1) Sum of squares between microcosms $S_{AB} = \sum_{ij} c \left(\bar{y}_{ij.} - \bar{y}_{...} \right)^2$

(2) Sum of squares between populations $S_A = \sum_i bc \left(\bar{y}_{i..} - \bar{y}_{...} \right)^2$

(3) Sum of squares in a population $S_{B(A)} = S_{AB} - S_A$

(4) Sum of squares of time and population $S_{AC} = \sum_{ik} b \left(\bar{y}_{i.k} - \bar{y}_{...} \right)^2$

(5) Sum of squares between times $S_C = \sum_k ab \left(\bar{y}_{..k} - \bar{y}_{...} \right)^2$

(6) Sum of squares of time × population $S_{A \times C} = S_{AC} - S_A - S_C$

(7) Total sum of squares $S_T = \sum_{ijk} \left(\bar{y}_{ijk} - \bar{y}_{...} \right)^2$

(8) Sum of squared error $S_E = S_T - S_{AB} - S_C - S_{A \times C}$

- Procedure (3): Calculate the degrees of freedom using the following expressions (1–7). Here, the flexibility is defined as various statistics. For example, the flexibility of the observation data with sample size n $(x_1, x_2, \ldots \ldots, x_n)$ becomes n:

(1) $\nu_A = a - 1$

(2) $\nu_{B(A)} = a(b - 1)$

(3) $\nu_{AB} = ab - 1$

(4) $\nu_c = c - 1$

(5) $\nu_{A \times C} = (a - 1)(c - 1)$

(6) $\nu_E = abc - 1 - \nu_{AB} - \nu_C - \nu_{A \times C}$

(7) $\nu_T = \nu_A + \nu_{B(A)} + \nu_C + \nu_{A \times C} + \nu_E$

- Procedure (4): Calculate the unbiased variance using the following expressions (1–5), where (3) is the value of the sum of squares divided by the degree of freedom.

Table 6.2 Table of branching-type ANOVA (ex. (2)-② means procedure (2)-②)

Factor	Sum of square	Flexibility	Unbiased variance	F-ratio	F value $100\alpha\%$ point*
Group	S_A (2)-②	ν_A (3)-①	V_A (3)-⑨	F_A (4)-①	$F(\nu_A,\ \nu_{B(A)},\ \alpha)$
Individual (group)	$S_{B(A)}$ (2)-③	$\nu_{B(A)}$ (3)-②	$V_{B(A)}$ (3)-⑧	$F_{B(A)}$ (4)-②	$F(\nu_{B(A)},\ \nu_E,\ \alpha)$
Time point	S_C (2)-⑤	ν_C (3)-④	V_C (3)-⑪	F_C (4)-③	$F(\nu_C,\ \nu_E,\ \alpha)$
group×time point	$S_{A \times C}$ (2)-⑥	$\nu_{A \times C}$ (3)-⑤	$V_{A \times C}$ (3)-⑩	$F_{A \times C}$ (4)-④	$F(\nu_{A \times C},\ \nu_E,\ \alpha)$
Residual error	S_E (2)-⑧	ν_E (3)-⑥	V_E (3)-⑫		
Sum	S_T (2)-⑦	ν_T (3)-⑦			

(1) $V_{B(A)} = S_{B(A)}/\nu_{B(A)}$
(2) $V_A = S_A/\nu_A$
(3) $V_{A \times C} = S_{A \times C}/\nu_{A \times C}$
(4) $V_C = S_C/\nu_C$
(5) $V_E = S_E/\nu_E$

- Procedure (5): Calculate an alpha statistic (i.e., level of significance) using the next expressions (1–4). The following values were obtained according to the F-distribution. Here, the F-distribution is a consecutive probability distribution used in statistics and probability theory. It is used as the distribution following the null hypothesis in the F-test and is applied to the analysis of variance.

(1) $F_A = V_A/V_{B(A)}$
(2) $F_{B(A)} = V_{B(A)}/V_E$
(3) $F_C = V_C/V_E$
(4) $F_{A \times C} = V_{A \times C}/V_E$

- Procedure (6): Determine the alpha statistic, α, and calculate the F-statistics (νA, νE, and α).

- Procedure (7): A significant difference is determined in comparison with the control system if the F-ratio is larger than the % point (of the F-distribution). Table 6.2 lists the calculated results of the analysis of variance. In the microcosm test method, $\alpha = 0.05$ is used to determine whether there is a significant difference (5% is generally adopted as the level of significance in this technique).

When the F-ratio of a point of group and crowd X** is larger than the F-statistic, it is determined that the addition of the chemical substance has an influence on the microcosm

**The factor is called the point of group X with a group and the interaction of the point in time, and the change in the reaction with the point in time differs among groups. For example, the value in the control system is stable, but, in the addition system, values may increase or decrease. Therefore, when there is a significant difference at the point of group X, there was an influence, even if the numerical mean was the same all day.

6.3.2   Calculation Example

The nonylphenol addition test of the microcosm provides numerical values, such as those shown in Table 4.3, and the results of the branching-type ANOVA are presented in this chapter. An example of the calculations used to analyze the influence of nonylphenol, which uses the P and R in the microcosm, is also shown below. The experimental conditions in a normal microcosm test (refer to Sect. 3.1) are shown in Table 6.3.

Based on a dataset (described in Table 6.4) of the measurement period 14 days after the addition of nonylphenol that was repeated three times (three replicates being

Table 6.3   Experimental condition of nonylphenol addition

Temperature	25
Illuminance	2400 lx
Cultivation period	30 days
Nonylphenol addition	16 days after cultivation started
Nonylphenol concentration	0.1 mg/L, 1 mg/L

Table 6.4   Measurement value of respiration amount in control system and nonylphenol added system

Days	Control			0.1 mg/L added			1 mg/L added		
	1	2	3	1	2	3	1	2	3
Before	0.44	0.66	0.61	0.45	0.40	0.44	0.55	0.37	0.70
Added 1	0.46	0.75	0.58	0.47	0.44	0.40	0.86	0.74	0.91
2	0.42	0.65	0.61	0.42	0.42	0.47	1.30	1.42	1.47
3	0.48	0.53	0.56	0.43	0.43	0.47	0.73	0.79	0.95
4	0.63	0.55	0.63	0.55	0.48	0.48	0.70	0.61	0.80
5	0.65	0.58	0.65	0.54	0.41	0.48	0.61	0.51	0.74
6	0.53	0.66	0.62	0.46	0.41	0.48	0.50	0.45	0.67
7	0.47	0.50	0.62	0.35	0.40	0.48	0.45	0.40	0.64
8	0.43	0.49	0.57	0.39	0.40	0.48	0.44	0.43	0.62
9	0.38	0.49	0.52	0.35	0.36	0.48	0.38	0.49	0.58
10	0.33	0.44	0.58	0.33	0.35	0.48	0.36	0.51	0.56
11	0.30	0.46	0.59	0.29	0.32	0.50	0.37	0.46	0.58
12	0.36	0.37	0.61	0.35	0.30	0.52	0.38	0.48	0.60
13	0.40	0.33	0.53	0.40	0.35	0.53	0.39	0.45	0.60
14	0.48	0.33	0.65	0.40	0.41	0.60	0.43	0.49	0.65

(Output value)

needed for statistical analyses), a calculation procedure and the results analyzed using a branching-type ANOVA are shown. For example, the analytical procedure for the no-effect scenario (0.1 mg/L of added nonylphenol) and the case in which there were effects (1 mg/L of added nonylphenol) are indicated below. The process of practically calculating the effects (in this case, the lack of effect) of the 0.1 mg/L addition of nonylphenol to the system is described. The "addition system" in the following procedures refers to the addition of 0.1 mg/L of nonylphenol. The formula is calculated according to the expressions listed above.

By this method, it becomes the number of populations, $i = 1, 2$ (1 represents the control system and 2 represents the addition system), the number of experiments, $j = 1, 2, 3$, and the time mark, $k = 1, 2, 3, 14$.

– Procedure (1)

(1) Mean of the control system 14 days after the addition in the first replicate $(i = 1, j = 1)$:

$\bar{y}_{12\cdot} = (0.46 + 0.42 + 0.48 + 0.63 + 0.65 + 0.53 + 0.47 + 0.43 + 0.38 + 0.33 + 0.3 + 0.36 + 0.4 + 0.48) \div 14 = 0.451$

Mean of the control system 14 days after addition in the second replicate $(i = 1, j = 2)$:

$\bar{y}_{12\cdot} = $ (the sum in the control system 14 days after addition in the second replicate) $\div 14 = 0.509$

As above, calculate in order, $\bar{y}_{13\cdot}$ $(i = 1, j = 3) = 0.594$, $\bar{y}_{21\cdot}$ $(i = 2, j = 1) = 0.410$, $\bar{y}_{22\cdot}$ $(i = 2, j = 2) = 0.392$, $\bar{y}_{23\cdot}$ $(i = 2, j = 3) = 0.491$. ($i = 1$ represents the control system, $i = 2$ represents the addition system, and j represents the repetitions.)

(2) The mean in each population, namely, the mean of the three replicates of the control system data 14 days after addition:

$\bar{y}_{1\cdot\cdot} = $ (the sum of the first replicate of the control system data 14 days after addition + the sum of the second replicate of the control system data 14 days after addition + the sum of the third replicate of the control system data 14 days after addition) $\div (14 \times 3) = 0.518\cdots$.

$\bar{y}_{2\cdot\cdot} = $ (the sum of the first replicate of the addition system data 14 days after addition + the sum of the second replicate of the addition system data 14 days after addition + the sum of the third replicate of the addition system data 14 days after addition) $\div (14 \times 3) = 0.431$.

(3) Mean for each time and population:

Control system:

$\bar{y}_{1\cdot1} = (0.46 + 0.75 + 0.58) \div 3 = 0.597$, $\bar{y}_{1\cdot2} = $ (the sum of the second-day data of the three replicates of the control system) $\div 3 = 0.559$, $\cdots\cdots$, $\bar{y}_{1\cdot14}$ = (the sum of the 14th-day data of the three replicates of the control system) $\div 3 = 0.486$.

Addition system:

$\bar{y}_{2\cdot1} = (0.47 + 0.44 + 0.40) \div 3 = 0.439$, $\bar{y}_{2\cdot2} = $ (the sum of the second-day data of the three replicates of the addition system) $\div 3 = 0.439$, $\cdots\cdots$, $\bar{y}_{2\cdot14}$

= (the sum of the 14th-day data of the three replicates of the addition system)
$\div 3 = 0.471$.

(4) Mean in each time:

$\bar{y}_{..1} = (0.46 + 0.75 + 0.58 + 0.47 + 0.44 + 0.4) \div 6 = 0.518$, $\bar{y}_{..2} =$ (the sum of the second-day data in the control system + the sum of the second-day data in the addition system) $\div 6 = 0.499, \cdots\cdots$, $\bar{y}_{..14} =$ (the sum of the 14th-day data in the control system + the sum of the 14th-day data in the addition system) $\div 6 = 0.479$.

(5) Grand mean:

$\bar{y}_{...} =$ (mean of all data in the control system and addition system) $= 0.474$.

- Procedure (2)

(1) $S_{AB} = 14 \times \{(0.451-0.474)^2 + (0.509-0.474)^2 + \cdots \cdots (0.491-0.474)^2\} = 0.38$.

(2) $S_A = 3 \times 14 \times \{(0.518-0.474)^2 + (0.431-0.474)^2\} = 0.159$.

(3) $S_{B(A)} = (0.382-0.159) = 0.223$.

(4) $S_{AC} = 3 \times \{(0.597-0.474)^2 + (0.559-0.474)^2 + \cdots \cdots (0.471-0.474)^2\} = 0.406$.

(5) $Sc = 2 \times 3 \times \{(0.518-0.474)^2 + (0.499-0.474)^2 + \cdots \cdots (0.479-0.474)^2\} = 0.199$.

(6) $S_{A \times C} = 0.406-0.159-0.199 = 0.0479$.

(7) $S_T = \{(0.46-0.474)^2 + (0.42-0.474)^2 + \cdots \cdot (0.60-0.474)^2\} = 0.872$.

(8) $S_E = 0.872-0.382-0.199-0.047 = 0.242$.

- Procedure (3)

(1) $\nu_A = 2-1 = 1$

(2) $\nu_{B(A)} = 2(3-1) = 4$

(3) $\nu_{AB} = 2 \times 3-1 = 5$

(4) $\nu_C = 14-1 = 13$

(5) $\nu_{A \times C} = (2-1)(14-1) = 13$

(6) $\nu_E = 2 \times 3 \times 14-1-5-13-13 = 52$

(7) $\nu_T = 1 + 4 + 13 + 13 + 52 = 83$

- Procedure (4)

(1) $V_{B(A)} = 0.223 / 4 = 0.0559$

(2) $V_A = 0.159 / 1 = 0.159$

(3) $V_{A \times C} = 0.047 / 13 = 0.00368$

(4) $V_C = 0.199 / 13 = 0.0153$

(5) $V_E = 0.242 / 52 = 0.00465$

- Procedure (5).

(1) $F_A = 0.159 / 0.0559 = 2.84$.

(2) $F_{B(A)} = 0.0559 / 0.00465 = 12.00$.

(3) $F_C = 0.0153 / 0.00465 = 3.29$.

Table 6.5 Table of branching-type ANOVA (example of no-effect, nonylphenol 0.1 mg/l)

Factor	Sum of square	Flexibility	Unbiased variance	F-ratio	F value ($\alpha = 0.05$) *
Group	0.16 (2)-②	1 (3)-①	0.16 (3)-⑨	2.84 (4)-①	7.71
Individual (group)	0.22 (2)-③	4 (3)-②	0.06 (3)-⑧	12.00 (4)-②	2.55
Time point	0.19 (2)-⑤	13 (3)-④	0.02 (3)-⑪	3.29 (4)-③	1.91
group×time point	0.05 (2)-⑥	13 (3)-⑤	0.004 (3)-⑩	0.79 (4)-④	1.91
Residual error	0.24 (2)-⑧	52 (3)-⑥	0.005 (3)-⑫		
Sum	0.87 (2)-⑦	83 (3)-⑦			

Table 6.6 Table of branching-type ANOVA (example of no-effect, nonylphenol 1 mg/l)

Factor	Sum of square	Flexibility	Unbiased variance	F-ratio	F value ($\alpha = 0.05$)
Group	0.27	1	0.27	2.65	7.71
Individual (group)	0.41	4	0.10	22.79	2.55
Time point	1.67	13	0.13	28.94	1.91
group×time point	1.04	13	0.08	18.04	1.91
Residual error	0.23	52	0.004		
Sum	3.62	83			

(4) $F_{A \times C} = 0.00368 / 0.00465 = 0.79$.

- Procedure (6): Fix a level of significance for $\alpha = 0.05$ and calculate the F-statistic corresponding to each degree of freedom.
- Procedure (7): Summarize the aforementioned calculation results in an analysis of variance list (Table 6.5). Based on this list, judge the factor as significant if the F-ratio of a population or population x time is greater than the F-statistic.

One example from Table 6.4 (0.1 mg/L nonylphenol addition system) reveals that the F-ratio is smaller than the F-statistic ($\alpha = 0.05$) in both the population (F-ratio < F-statistic, 2.85 < 7.71) and population x time (F-ratio < F-statistic, 0.79 < 1.91). As there is no significant difference, the system with added nonylphenol is considered a "no-effect" system (Table 6.6).

In another example, from Table 4.5 (1 mg/L nonylphenol addition system), a significant difference in the populations was estimated due to the F-ratio exceeding the F-statistic ($\alpha = 0.05$) of population x time (F-ratio > F-statistic, 18.04 > 1.91), although the F-ratio was smaller than the F-statistic ($\alpha = 0.05$) of the population (F-ratio < F-statistic, 2.65 < 7.71). It was estimated that this system was affected

Table 6.7   Means of each factor in branching-type ANOVA

	P/R	P	R	judgement
Case 1 :	○	○	○	no influence
Case 2 :	○	×	×	influence
Case 3 :	×	×	○	influence
Case 4 :	×	○	×	influence
Case 5 :	×	×	×	influence

＊○ : without significant difference, × : with significant difference

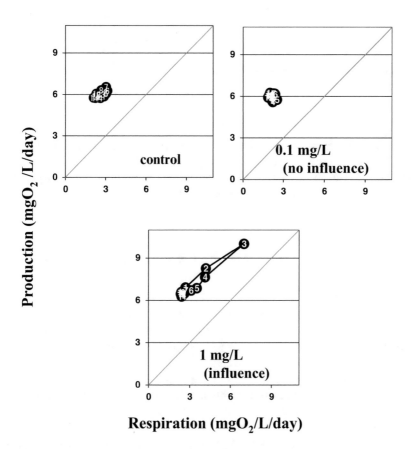

Fig. 6.6 Time course of P/R ratio in nonylphenol addition

because the respiration in the addition N-system differed from that in the control system (Table 6.7).

The result obtained from an analysis of the influence of nonylphenol on the P/R ratio by a branching-type ANOVA is shown in Fig. 6.6. From this example, it can be

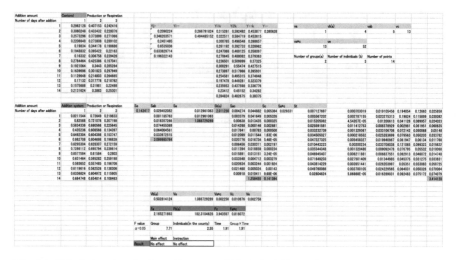

Fig. 6.7 Calculation sheet for branching-type ANOVA by MS Excel

understood that there was no ecological influence from nonylphenol on the microcosm N-system, based on the lack of a significant difference at 0.1 mg/L of added nonylphenol. However, there was an influence at 1 mg/L of added nonylphenol, as shown by the significant difference in the population.

6.3.3 Use of Microsoft Excel for Analyzing the Microcosm N-System

The calculation method for a branching-type ANOVA using MS Excel is shown in Figs. 6.7, 6.8, and 6.9. Figure 6.7 shows the outline of the calculation process, while Fig. 6.8 shows how and where to input production and respiration amounts, and Fig. 6.9 shows the displayed results. It can be examined instantly by using the calculation sheet presented below. In other words, whether there is a significant difference can be easily determined by inputting P and R data into the MS Excel calculation sheet, which calculates the relevant statistics automatically.

When the calculation program of the branching-type ANOVA built in MS Excel is opened, the branching-type ANOVA sheet shown in Fig. 6.7 is displayed. The measured amount of production or respiration in the control system is input into the upper yellow frame of the input cell (Fig. 6.8) for comparison with the amount of production and respiration provided by the DO logger on the left of this screen. When a value is input, whether there is a significant difference in the value of the main effect and the interaction is judged automatically, and the presence or

Addition amount	Contorol	Production or Respiration	
Number of days after addition	1	2	3
1	0.2962128	0.407153	0.242416
2	0.3060248	0.433432	0.228076
3	0.2573296	0.373899	0.271066
4	0.2206648	0.273808	0.289102
5	0.19834	0.344178	0.198686
6	0.1948832	0.395422	0.22163
7	0.16332	0.306758	0.239426
8	0.2784464	0.425386	0.197041
9	0.1921064	0.3443	0.285284
10	0.1639696	0.301823	0.297949
11	0.1129648	0.214803	0.264665
12	0.17132	0.317778	0.218762
13	0.1575688	0.31981	0.22498
14	0.2131624	0.3883	0.25301

Addition amount	Addition system	Production or Respiration	
Number of days after addition	1	2	3
1	0.8211544	0.73649	0.219833
2	0.82588	0.721078	0.207198
3	0.5834336	0.680566	0.225645
4	0.435236	0.608566	0.134397
5	0.6492304	0.604398	0.153747
6	0.662708	0.604698	0.196839
7	0.6295304	0.628307	0.272159
8	0.7286112	0.695794	0.239814
9	0.6577584	0.61584	0.2803
10	0.631464	0.595282	0.259198
11	0.593652	0.557485	0.196706
12	0.6119616	0.562526	0.138306
13	0.6336824	0.604972	0.115905
14	0.684748	0.654814	0.108463

Fig. 6.8   Enter of production and respiration amount on MS Excel sheet

	Fa	Fb(a)		Fc	Fa*c
	2.165271893		182.3104628	3.943597	0.818072
F value	Group	Individuals(in the county)		Time	Group × Time
α=0.05	7.71	2.55		1.91	1.91
	Main effect	Inetraction			
Result	No effect	No effect			

Fig. 6.9   Calculation results of branching-type ANOVA on MS Excel sheet

absence of an influence is displayed in the lower section of the MS Excel screen (Fig. 6.9).

6.3.4   Standard of Environmental Impact Assessment on the Microcosm N-System

The final criteria of whether or not a given substance (i.e., test material) had an impact on the microcosm are as follows:

Table 6.8 Standard table for influence assessment

	P/R	P	R	Judgment
Case 1	○	○	○	No influence
Case 2	○	X	X	Influence
Case 3	X	X	○	Influence
Case 4	X	○	X	Influence
Case 5	X	X	X	Influence

○:without significant difference, X:with significant difference

- The endpoints are (1) respiration amount (R) and production amount (P) and (2) the P/R ratio; the final criteria for a judgment of "no-effect" are that there are no significant differences in the values of R and that the P/R ratios converge to 1.
- Judgment parameters are (1) determined on the 30th day after cultivation began, and (2) changes are recorded in the period of 14 days after a test material is added (in case a test agent was added 16 days after microcosm cultivation began).

In other words, whether there is a significant difference is determined, and such determinations are shown in Table 6.8. As for the P/R ratio, P and R are statistically analyzed using a branching-type ANOVA, and the influence of collating with this criteria table is estimated. Additionally, the branching-type ANOVA described above was used to assess the microcosm N-system.

Appendix for Statistical Analyses

1. Statistical Testing

Statistical testing of hypotheses is performed using specimens, with which the verification of various hypotheses regarding a population is performed. A concept of statistical testing is shown in Fig. 6.10).

◎ Statistical testing: Comparison of the differences among multiple specimens is performed, and it is estimated whether it may be said that there is a difference from the result to the population corresponding to each specimen.【Example】The mean of specimen B is compared with the mean of specimen A, and it can be said that there is a difference between the mean of population A and the mean of population B from the result.

◎ Significant probability, confidence interval: The probability is the possibility that the result for a real specimen will not differ between populations (i.e., when the null hypothesis is supported). It can be said that there is a significant difference when this probability is less than 5% (i.e., the null hypothesis of no difference between populations is rejected).

◎ Why must verification be performed? When data for the entire population are available for all cases, verification is not necessary. In many cases, the data available from specimens represent only small portions of the population being investigated.

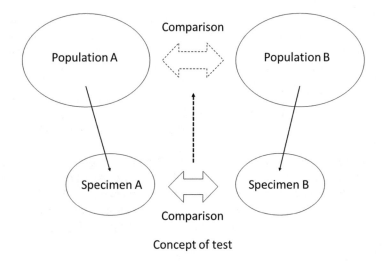

Concept of test

Fig. 6.10 Concept of statistical testing

However, the question being investigated here is not "is there a difference between *specimens*?" but rather "is there a difference between *populations*?", and it must be answered based on conclusions derived from a limited number of specimens. Therefore, when the difference between specimens is too small, it can be concluded that there is no significant difference between the populations, and when the difference between specimens is large, to some extent, it can be said that there is a significant difference between the populations. In that case, the significant difference will be expressed as a significant difference in the probability ($\bullet$%), and this probability (level of significance) indicates the possibility of incorrectly arriving at the conclusion that there is a significant difference.

◎ Rearranging of the main term for the statistical test:

①Specimen: Sample size of the ※ specimens from some population of the target group to be investigated.
②Population: The large (entire) group.
③Random sampling: Method of choosing a random specimen from within a population.
※Probability sample, random sample: The specimens that were randomly selected.
④Parameters: Generic names, such as the mean (population mean) of the population, dispersion (population dispersion), and the standard deviation (population standard deviation).
⑤Statistics: Generic names, such as the mean (sample mean) of the specimen, dispersion (sample dispersion), and the standard deviation (sample standard deviation).
⑥Regular population: Population distribution of the regular model.

※Normal distribution: The majority of experimental values have a central tendency, with values increasing and decreasing away from a central value, presenting a symmetrical bell curve. If samples are normally distributed, statistical analysis is possible. However, if the mean is skewed, some amount of error is possible.

⑦Null hypothesis: A hypotheses of no difference between the populations being compared. To determine whether or not there are differences between populations, a null hypothesis of "no difference" is used. An alternative hypothesis (the hypothesis that there is a difference in) is adopted if the null hypothesis is rejected.

⑧Confidence interval significant probability: The probability (possibility) that the null hypothesis is supported from the results produced from a real specimen. It is said that there is a significant difference when this probability is less than 5% (i.e., the null hypothesis is rejected as almost impossible). It is denoted as "α" or "p" and expressed as either a probability (p = 0.01) or a percentage (p = 1%).

2. Difference Between Test Types and Their Uses

Tests are divided into analyses of variance (ANOVA) for judging the difference among groups, the Student's t-test for judging the difference between two groups, and multivariate techniques. The t-test is used to examine whether there is a difference between the population means between two independent groups, such as A and B. For three groups such as A, B, and C, for which μ represents the mean, it would be a serious mistake to simply repeat the t-test as below. In this case, multivariate techniques must be used.

- Between A and B, is there a difference in the value? ($\mu_A = \mu_B$)
- Between A and C, is there a difference in the value? ($\mu_A = \mu_C$)
- Between B and C, is there a difference in the value? ($\mu_B = \mu_C$)

- Why must the t-test not simply be repeated for multiple groups?

In the case of an examination with 5% confidence intervals, the probability (of a type I error) to dismiss the null hypothesis, H_0, by mistake is 0.05, assuming that the H_0 of $\mu A = \mu B$ is true. When separate data analyses are performed for various problems associated with a certain sample, three H_0 will be considered the truth when analyses are performed three times, and the probability of type I errors occurring somewhere in the article will become unexpectedly high, at $1 - (0.95 \times 0.95 \times 0.95) = 0.14$. When the same analyses are performed 10 times, the probability of a type I error becomes 0.40. Furthermore, a more serious problem occurs when different tests are based on the same data. For example, if a t-test of A is performed among the three groups, A, B, and C, comparisons are performed between A vs. B, B vs. C, and C vs. A. Unfortunately, a sample mean of A will then be higher than the true average; in the case of A vs. B and A vs. C, H_0 becomes easy to reject.

Fig. 6.11 Notice matter of the multiplex nature by the repetition of the multiple comparison

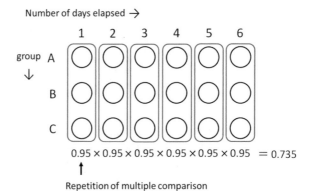

$$0.95 \times 0.95 \times 0.95 \times 0.95 \times 0.95 \times 0.95 = 0.735$$

Repetition of multiple comparison

Fig. 6.12 Concept of branching-type ANOVA

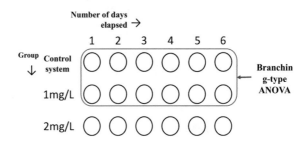

Therefore, when three H_0 are considered to be the truth, the probability of a type I error is greater than $1-0.95^3 = 0.14$, and it will be unpredictable.

In this way, when the t-test is simply repeated:

- The probability of a type I error throughout the analysis increases.
- When data are not independent, the probability of the type I error becomes unknown.

In this case, it is necessary to use multivariate techniques. However, in the daily operation of the microcosm and after having examined groups A, B, and C on the first day by multiple comparisons, on the subsequent days, the test will be repeated. Therefore, it may be said that this is statistically a mistake because the probability of rejecting the null hypothesis, H_0 ($\mu A = \mu B = \mu C$), for groups A, B, and C increases with the repetition of the t-test. A notice matter of the multiplex nature by the repetition of the multiple comparison is shown in Fig. 6.11.

The branching-type ANOVA can examine the control system and the addition system (1 mg/L) for all measured days together. Here, all groups could be examined at the same time, but it was decided that a closed testing order would be used because it is not known whether there is any influence from the addition of 1 mg/L or 2 mg/L. A concept of branching-type ANOVA is shown in Fig. 6.12.

A method of using a closed testing order is recommended in the dose-response related examination to resolve the problems associated with the multiplex nature

Fig. 6.13 Way of thinking
of the closed testing order

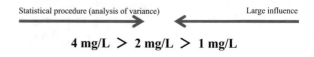

**Determine the importance of the test hypothesis
and decide the priority of the test.**

Statistical procedure (analysis of variance)   Large influence

4 mg/L > 2 mg/L > 1 mg/L

when examining every 1 mg/L and 2 mg/L of addition concentration. The importance for each hypothesis is set, and a closed testing order is a technique that allows for sequential examination by importance. When an insignificant result is provided, verification ends there. In the closed testing order, the entire examination and an equal level of significance can be used for verification, or preset levels of importance may be substituted beforehand. In other words, in the microcosm system, it is thought that higher addition concentrations result in greater effects, so the addition concentration is examined in descending order. Verification is terminated at the concentration with no significant difference, and this is assumed to be the maximum NOEC. The Way of thinking of the closed testing order is shown in Fig. 6.13.

Literature Cited

Bennet A, Bogorad L. Complementary chromatic adaptation in a filamentous blue-green alga. J Cell Biol. 1973;58:419–35.

Burnison BK. Modified dimethyl sulfoxide (DMSO) extraction for chlorophyll analysis of phytoplankton. Can J Fish Aquat Sci. 1980;37:729–33.

Ichihara K. Statistics of the bioscience – practice theory for right and inflects. Nankoudo; 1990. p. 378.

Japanese Red Cross Toyota College of Nursing. Basics of statistical analysis to make use of in nursing, Toyota. 2005. 136pp.

Kakidume T. About the dose-response-related examination in the clinical trial. Lect Note Inst Math Anal. 2008;1603:1–10.

Kataya A, Matsufuji T. Introduction to environmental statistics -viewpoint, summary of environment data. Ohmsha; 2003. p. 164.

Ono S. The basics of analysis of variance to understand by reading. 2nd ed. Tokyo. website; 2003.

Patel A, et al. Purification and characterization of C-Phycocyanin from cyanobacterial species of marine and freshwater habitat. Protein Exp Purif. 2005;40:248–55.

Wellburn AR. The spectral determination of chlorophylls a and b, as well as total carotenoids, using various solvents with spectrophotometers of different resolution. J Plant Physiol. 1994;144:307–13.

WHO. Principles for evaluating health risks to reproductions associated with exposure to chemicals. Environmental Health Criteria 225, Geneva: WHO; 2001.

Chapter 7
Example Assessments of the Microcosm N-System

Kazuhito Murakami, Kunihiko Kakazu, Kakeru Ruike, and Ken-ichi Shibata

Abstract In the example assessment, not only chemicals but also metals, radiation, and nonindigenous microorganisms were used as test materials. Estimations of the impacts of 106 examples of test materials (6 of surfactants, 3 of germicides, 14 of herbicides, 12 of pesticides, 2 of endocrine-disrupting chemicals, 1 of an antibiotic, 1 of an algal toxin, 6 of organic matter and solvents, 2 of nutrients, 15 of metals, 2 of radiation, 6 of microbial pesticides, 8 of genetically modified bacteria, 13 incorporating biomanipulation, and 2 involving climate change) are described in this chapter. It was made clear by the results of these tests that it is sufficient to analyze the P/R ratio with a branching-type ANOVA in the microcosm test. However, the characteristics of the constituent organisms were also evaluated, and the analyses are explained here.

7.1 Surfactants

The stability of aquatic environments in Japan has been improved, but many problems still remain in need of urgent resolution. One such problem is that great quantities of surfactants flow directly into aquatic environments with gray water, deteriorating the aquatic ecosystem, even though there is little problem in areas with waste management facilities because sewage is effectively treated. To assess the environmental effects of surfactants at the ecosystem level, it is necessary to analyze the mechanisms of community stability under the natural conditions present in aquatic environments because natural ecosystems are extremely complex and are

K. Murakami (✉)
Chiba Institute of Technology, Narashino, Chiba, Japan
e-mail: kazuhito.murakami@p.chibakoudai.jp

K. Kakazu · K. Ruike
University of Tsukuba/Foundation for Advancement of International Science, Tsukuba, Ibaraki, Japan
e-mail: kakazu@fais.or.jp; ruike@fais.or.jp

K. Shibata
Yokohama National University, Yokohama, Kanagawa, Japan
e-mail: shibata-kenichi-ym@ynu.jp

© Springer Nature Singapore Pte Ltd. 2020
Y. Inamori (ed.), *Microcosm Manual for Environmental Impact Risk Assessment*,
https://doi.org/10.1007/978-981-13-6798-4_7

exposed to unpredictable environmental factors. Instead of working with a natural ecosystem, it is convenient to use microcosms consisting of biotic and abiotic factors that originated from a natural ecosystem because they allow for both biological simplicity and replication (Beyers 1963; Cook 1967; Margalef 1969; Kawabata et al. 1978). In this section, the effects of surfactants on the aquatic ecosystem and the biodegradability of surfactants are described.

7.1.1 Linear Alkylbenzene Sulfonate (LAS)

A microcosm was used to assess the effects of the anionic surfactant, linear alkylbenzene sulfonate (LAS), on an aquatic ecosystem. The addition concentration of LAS was adjusted to 1, 5, and 10 mg/L and was added on the 16th day of the stable state into the microcosm. Linear alkylbenzene sulfonate supplied to this experiment was composed of a sodium linear-dodecyl benzene sulfonate standard (C12) and LAS at a purity of more than 99%.

The endpoints were the population density (structural parameter) and DO (functional parameter), and the population was measured by an optical microscope and counted from the beginning of culturing on days 0, 2, 4, 7, 14, 16, 18, 20, 23, and 30, and it was evaluated from the results of B16–30 (days 16–30), which was the ratio of the abundance and population density (N_{30}) on the 30th day. The DO was measured continuously from the 16th day onward, and the P/R ratio was calculated from the amounts of production (P) and respiration (R). Using the structural parameter, the impact of LAS was evaluated in comparison with the control system according to the behaviors of the microorganisms in the microcosm on each of the 14 days after the addition of LAS for which measurements were taken. The population that did not change relative to the control system was maintained at a 1 mg/L surfactant load and recovered to almost the same population density, while there was a decrease in the density of microanimals in the microcosm with the 5 mg/L load. In the microcosm with a 10 mg/L load of LAS, *Cyclidium glaucoma* and *Philodina erythrophthalma* disappeared, and the populations of *Lecane* sp., *Aeolosoma hemprichi*, and *Tolypothrix* sp. were stable at low densities, and the microcosm ecosystem did not collapse.

As a result of the observed changes in the structural parameter, the m-NOEC (microcosm maximum effect-free concentration) of LAS was estimated as 1 mg/L. While the NOEC of LAS for *Daphnia magna* in a single-species examination was 1180–3250 µg/L and the NOEC for algae was 400–18,000 µg/L, it was thought that the m-NOEC exhibited slight unevenness. It was shown that the evaluations of the microcosm test using the structural parameter (i.e., population density) and the functional parameter (i.e., the P/R ratio) were consistent. The microcosm test is an effective tool for assessing the effects of surfactants on ecosystems because it allows for the evaluation of the effects of surfactants from the viewpoint of the interactions among microorganisms, material cycling, and energy flow. Based on these characteristics, the microcosm test is a useful method of performing environmental assessments that can reflect the behavior of natural aquatic ecosystems.

Nearly the same effects and degradation were observed when LAS was added to the scaled-up microcosm, which included the same microbiota as those in the culture medium of the flask-sized microcosm. This microcosm was scaled up to a 10 L glass jar (depth, 30 cm; surface area, 490 cm^2) containing 7 L (depth, 16 cm) of TP medium. In this experiment, concentrations of LAS, adenosine triphosphate (ATP), nutrients, DO, pH, oxidation reduction potential (ORP), chemical oxygen demands (COD), dissolved organic carbon (DOC), suspended solid (SS), and chlorophyll *a* (Chl.*a*) were also analyzed. When the initial concentration of LAS was 1.5 mg/L, changes in the abundance of all microorganisms were the same as those in the control system. At 2.5 mg/L, the abundance of *Cyclidium glaucoma* decreased for 2 days after the addition of LAS, but it slowly increased again. At 5.0 mg/L, *Cyclidium glaucoma* and *Tolypothrix* sp. were eliminated from the system. The abundance of *Philodina erythrophthalma* and *Aeolosoma hemprichi* decreased for 2 days after the addition of LAS, but it slowly increased again until equilibrating with that of the control system. The abundance of bacteria was ten times higher than that of the control on day 2 after the LAS addition. At 10 mg/L, the abundance of *Philodina erythrophthalma* and *Lecane* sp. decreased for 2 days after LAS addition and then slowly increased. *Aeolosoma hemprichi* was eliminated from the system. The abundance of bacteria was 100 times higher than in the control on days 2–7 after LAS addition. The effect of LAS on *Chlorella* sp. was not recognized in the LAS concentrations added in this experiment. The NOEC of LAS on the population density was less than 1.5 mg/L.

When the initial concentration of LAS was less than 2.5 mg/L, the concentration of ATP remained the same as in the control system (Fig. 7.1). At LAS concentrations of 5.0 mg/L and 10.0 mg/L, the concentration of ATP decreased, which corresponded to a decrease in the abundance of microorganisms, such as *Cyclidium glaucoma*, *Philodina erythrophthalma*, and *Tolypothrix* sp. It was found that ATP reflected the decrease in the population densities of the microcosm system. Evaluation on the basis of ATP showed that the NOEC of LAS was less than 2.5 mg/L. At initial LAS concentrations of less than 2.5 mg/L, the concentration of DO and the P/R ratio remained the same as those in the control system. At an LAS concentration of 5.0 mg/L, the concentration of DO and the P/R ratio decreased for 4 days after the addition of LAS, but, toward the end of the experiment, DO recovered to the same level as in the control system. At an LAS concentration of 10.0 mg/L, the concentration of DO decreased and converged to 3.6 mg/L; the P/R ratio also decreased for 4 days from 1.1 to −5.1, but, toward the end of the experiment, it recovered to the same level as in the control system. The NOEC of LAS was less than 2.5 mg/L when evaluated according to the concentration of DO and the P/R ratio. No effect of LAS on other measured parameters, such as pH, ORP, and nutrients, was observed at these initial LAS concentrations. These results demonstrate that the effect of LAS on microcosm population densities was simultaneously reflected by the concentrations of ATP and DO. Moreover, it was suggested that ATP and DO were sensitive to the changes in population densities caused by the addition of LAS and, thus, that using these parameters can allow for precise environmental assessments to be performed.

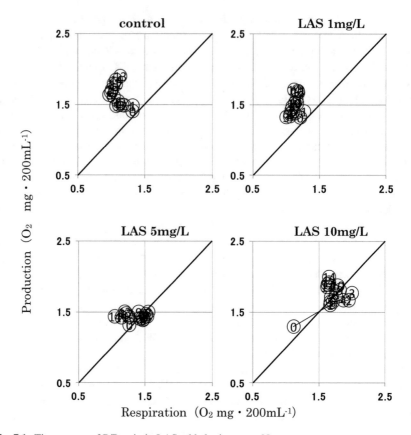

Fig. 7.1 Time course of P/R ratio in LAS-added microcosm N-system

7.1.2 Alcohol Ethoxylate (AE)

The addition concentration of alcohol ethoxylate (AE) was adjusted to 2, 10, 25, 50, and 100 mg/L and added on the 16th day of the stable state into the microcosm. The endpoints were abundance (structural parameter) and the DO concentration (functional parameter), and the population was measured using an optical microscope and counted the start of culturing on days 0, 2, 4, 7, 14, 16, 18, 20, 23, and 30, and it was evaluated from the results of B_{16-30} (days 16–30), which was the ratio of the abundance and the population density (N_{30}) on the 30th day. The concentration of DO was measured continuously from the 16th day onward, and the P/R ratio was calculated from the amount of production (P) and respiration (R). Using the structural parameter, a constant population was maintained for all microorganisms in the control system through the experimental period, and a constant population was likewise maintained at a 2 mg/L AE load. A decrease in microanimals was confirmed in both the 10 mg/L and 25 mg/L AE load systems until the 2nd and 4th days, respectively, but their populations had recovered to nearly the same as that in the control system after the 7th day. The populations of *Cyclidium glaucoma*, *Aeolosoma hemprichi*, *Philodina erythrophthalma*, and *Lecane* sp. were affected, and the fluctuations grew large with increasing addition concentrations of AE.

The functional parameter (i.e., the P/R ratio) remained constant as production and respiration remained the same in the control system throughout the measurement period; approximately the same behavior was observed even at an AE concentration of 2 mg/L (Fig. 7.2). Production and respiration volumes increased just after the addition of 10 mg/L of AE, and a change was observed in comparison with the control system. However, the behavior of the 10 mg/L AE load system was similar to that of the control system after the 2nd day (i.e., the system was only disturbed temporarily, and it recovered afterward). Production and respiration volumes suddenly increased just after the addition of 25 mg/L of AE, and a remarkable change was observed; the P/R ratio remained less than 1 for 7 days after the addition of AE. This continuous state of a P/R < 1 meant the reduction of food resources by overconsumption, and it was thought that a load of 25 mg/L of AE heightened the risk of ecosystem collapse in the microcosm because a risk for future ecosystem collapse was expressed by the constituent microorganisms. A system with a 25 mg/L load of AE exhibited strong disturbance and risked collapse, but its behavior was similar to that of the control system after the 12th day. There was also an observed difference in the influence of the culture period, and it was suggested that the pattern of influence was similar to that of the 10 mg/L AE load. The importance to strengthen including degree and periods from influence by the chemical substance addition to recovery was suggested.

The results of the P/R ratio and the behavior of the population were consistent. However, a change might occur in the behavior of the P/R ratio even if the population is maintained at the same level as in the control system, and it is thought that the behavior of the P/R ratio broadly reflects changes in the population. The utility of the risk evaluation using the P/R ratio as an evaluation index was also demonstrated. No influence was observed in either the P/R ratio or the population densities at an AE concentration of 2 mg/L, which was estimated as the m-NOEC as a result of this experiment. However, it was thought that there was a need for endpoint setting that considered resistance to change, the analysis of patterns of resiliency, and relaxation of the influence of an autoregulatory function for ecosystem disturbance in natural environments modeled by the microcosm.

7.1.3 Sodium Dodecyl Sulfate (SDS)

The addition concentration of sodium dodecyl sulfate (SDS) was adjusted to 4, 8, and 16 mg/L and was added on the 16th day of the stable state into the microcosm. The endpoints were abundance (structural parameter) and the concentration of DO (functional parameter), and the population was measured using an optical microscope and counted from the start of culturing on days 0, 2, 4, 7, 14, 16, 18, 20, 23, and 30. It was evaluated from the results of B_{16-30} (days 16–30), which was the ratio of abundance and the population density (N_{30}) on the 30th day.

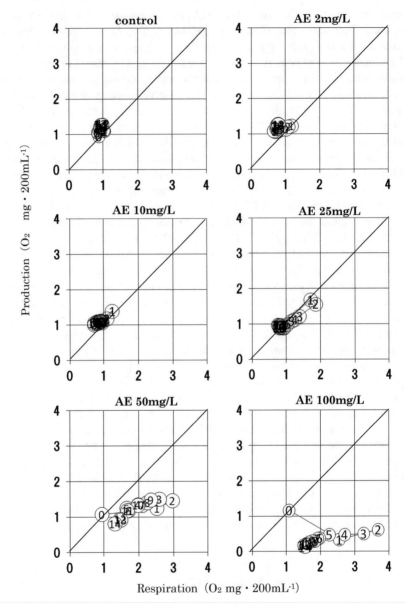

Fig. 7.2 Time course of P/R ratio in AE-added microcosm N-system

The concentration of DO was measured continuously from the 16th day onward, and the P/R ratio was calculated from the amount of production (P) and respiration (R). Using the structural parameter, the influence of SDS was evaluated in comparison with the control system based on the behaviors of microorganisms in the microcosm observed for each of the 14 days after the addition of SDS for each SDS load system. Increases in the abundance of *Lecane* sp. at the 4 mg/L and 8 mg/L SDS loads, a decrease in and recovery of *Tolypothrix* sp. at the 8 mg/L SDS load, and a decrease in the abundance of *Tolypothrix* sp., a decrease and recovery of *Cyclidium glaucoma*, and an increase in the abundance of *Lecane* sp. at the 16 mg/L SDS load were observed.

The influence of SDS was evaluated using the functional parameter (i.e., the P/R ratio) by statistically comparing the production and respiration volumes 14 days after the addition of the SDS load. Of the system after SDS was understood for life in a system as an example of a more complimentary analysis, and the toxicity became extinct, there was no difference between the addition and control systems, and the m-NOEC for SDS was estimated as 4 mg/L (Figs. 7.3 and 7.4). The respiration volume in the microcosm deviated from that of the control system after the 3rd day of SDS addition and was later restored in the 8 mg/L SDS load system. A change in the respiration volume was recognized on the next day following the addition of SDS and recovered afterward. However, a difference in the production volume between the control system and the experimental microcosm was produced at the end for the evaluation period, with the production volume being adversely affected in the 16 mg/L SDS load system.

A tendency toward an increase in the abundance of biota was observed, but it was able to judge it from 4 mg/L of SDS addition system (the m-NOEC) because there were no changes in the ecological balance of production and respiration. It was found that there was a negative influence on the population at an SDS concentration of 8 mg/L, but the density recovered to the control system level, whereas it could not recover in the 16 mg/L SDS load system. These results were similar to the predicted values from the mesocosm test of SDS according to the high correlation between the mean of the natural ecosystem (mesocosm test) (mean NOEC) and the m-NOEC. From these results, it can be estimated that the m-NOEC of SDS is 4 mg/L.

7.1.4   Trimethyldodecyl Ammonium Chloride (TMAC)

In the addition system with 10 mg/L of trimethyldodecyl ammonium chloride (TMAC), an increase in the abundance of the cyanophycean, *Tolypothrix* sp., and the ciliate, *Cyclidium glaucoma*, was observed, and the reverse was noted for the populations of the rotifer, *Philodina erythrophthalma*, and the oligochaete, *Aeolosoma hemprichi*. With the increase in the abundance of *Tolypothrix* sp. in particular, the color of the microcosm changed from green to brown. This is the reverse of the pattern reported for the decrease in the abundance of *Tolypothrix* sp. observed in the 16 mg/L SDS addition system.

A statistically significant difference between the addition and control systems was not recognized, and the m-NOEC was estimated as equal to 1 mg/L of TMAC (Fig. 7.5). There was a significant difference at the TMAC concentration of 10 mg/L. An impact assessment was performed by concentration setting of common ratio 2 in TMAC to calculate the m-NOEC. As significant differences emerged at the 0.2 mg/L TMAC load, the amount of production and respiration changed remarkably as the

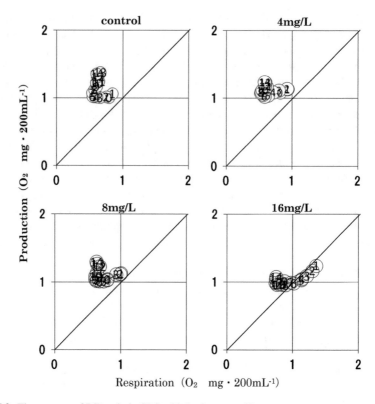

Fig. 7.3 Time course of P/R ratio in SDS-added microcosm N-system

concentration increased. From these results, it can be estimated that the m-NOEC of TMAC is 1 mg/L.

7.1.5 Soap

The addition concentration of soap was adjusted to 10, 30, 50, 100, 150, and 200 mg/ L, and it was added on the 16th day of the stable state into the microcosm. Soap supplied to this experiment was composed of sodium fatty acids (C8–C18). The endpoint was the population density (structural parameter) measured using an optical microscope and counted from the start of culturing on days 0, 2, 4, 7, 14, 16, 18, 20, 23, and 30, and it was evaluated from the results of B_{16-30} (days 16–30), which was the ratio of abundance and population density (N_{30}) on the 30th day. Using the structural parameter, the same density of microorganisms was maintained in the 10 mg/L and 30 mg/L soap addition systems as in the control system. *Cyclidium glaucoma* disappeared at concentrations of up to 50 mg/L 2 days after the addition of soap, but its population slowly recovered 4 days after the addition of soap and was

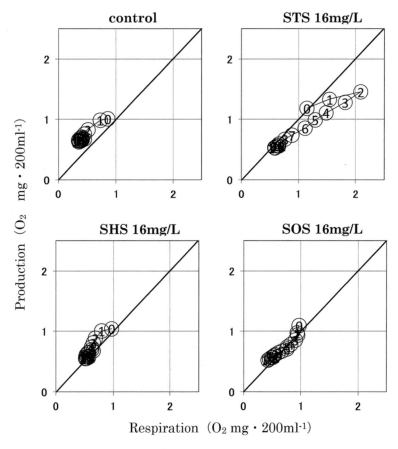

Fig. 7.4 Time course of P/R ratio in STS-, SHS-, and SOS-added microcosm N-system

maintained at a density of 1.0×10^2 N/mL. With the addition of 200 mg/L of soap, *Cyclidium glaucoma* disappeared from the system. It seems that the recovery of *Cyclidium glaucoma* was due to the degradation of complexed soap by bacteria. The density of *Tolypothrix* sp. slowly decreased to approximately 2.0×10^1 cm/mL 11 days after the addition of soap concentrations of up to 50 mg/L. The slow decrease in the density of *Tolypothrix* sp. was considered to have been driven by the light inhibition of soap scum. The densities of *Lecane* sp. and *Philodina erythrophthalma* slowly recovered in concentrations of up to 50 mg/L, ranging from 6.0×10^1 N/mL to 1.1×10^2 N/mL. The reason for the slow recoveries of *Lecane* sp. and *Philodina erythrophthalma* was assumed to be that they fed on bacteria, which increased in density due to the organic nature of complexed soap. With the addition of 200 mg/L of soap, the population of *Aeolosoma hemprichi* slowly decreased and finally disappeared from the system. Only *Chlorella* sp. was independent of the complexed soap at all concentrations. It follows from these

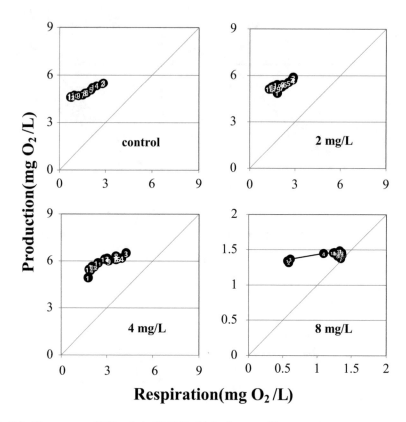

Fig. 7.5 Time course of P/R ratio in TMAC-added microcosm N-system

results that the populations of microorganisms in the system were not affected by complexed soap concentrations below 30 mg/L. As a result, it can be estimated that the m-NOEC of soap is less than 30 mg/L.

7.1.6 Complexed Soap

The addition concentration of complexed soap was adjusted to 10, 30, 50, 100, 150, and 200 mg/L, and it was added on the 16th day of the stable state into the microcosm. Complexed soap used in this experiment was composed of 58% soap and a lime soap-dispersing agent (LSDA). The endpoint was the population density (structural parameter) measured using an optical microscope and counted from the start of culturing on days 0, 2, 4, 7, 14, 16, 18, 20, 23, and 30, and it was evaluated from the results of B_{16-30} (days 16–30), which was the ratio of abundance and population density (N_{30}) on the 30th day. Using the structural parameter, the abundance of microorganisms maintained with the addition of 10 mg/L and 30 mg/L of complexed soap was the same as in the control system. *Cyclidium glaucoma* disappeared within 2 days of the addition of complexed soap concentrations of up to 50 mg/L, but slowly recovered in the 4 days after the addition of complexed soap and maintained a density of 1.2×10^2 N/mL. With the addition of 200 mg/L of complexed soap, *Cyclidium glaucoma* disappeared from the system. It seems that the recovery of *Cyclidium glaucoma* was due to the degradation of complexed soap by bacteria. The population density of *Tolypothrix* sp. slowly decreased to approximately 3.0×10^1 cm/mL with 15 days after the addition of complexed soap concentrations of up to 50 mg/L. The reason for the slow decrease in the population of *Tolypothrix* sp. was considered to be light inhibition by soap scum. The populations of *Lecane* sp. and *Philodina erythrophthalma* also slowly recovered at complexed soap concentrations of up to 50 mg/L, ranging from 3.0×10^1 N/mL to 8.9×10^1 N/mL. The reason for the slow recoveries of *Lecane* sp. and *Philodina erythrophthalm*a is considered to be their predation upon bacteria, which increased in abundance due to the organic nature of complexed soap. With the addition of 200 mg/L of complexed soap, the population of *Aeolosoma hemprichi* slowly decreased and finally disappeared from the system. Only *Chlorella* sp. was independent of complexed soap at all concentrations. It follows from these results that the population of microorganisms in the system was not affected by complexed soap concentrations below 30 mg/L. As a result, it can be estimated that the m-NOEC of complexed soap is less than 30 mg/L.

7.2 Germicides

Germicides (antiseptics) are antimicrobial substances, and they are applied to living tissues/skin to reduce the possibility of infection, sepsis, or putrefaction. Blast disease has been known as one of the serious blights on rice in Japan, and a preventative method of sterilizing seed rice with formalin was developed in the 1930s. Organomercuric agents were used widely after World War II, but due to another type of organic mercury-causing Minamata disease, of which the risk was noted in the 1960s, organomercuric agents were prohibited from use. For treating rice blast disease, blasticidin and kasugamycin were discovered and developed for agricultural use. Furthermore, some effective chemical agents for blast disease, such as dithiocarbamate in the 1960s and azole and benzimidazole in the 1970s, were developed. New chemical agents, including quinone outside inhibitors (QoI), have even been developed recently.

7.2.1 *2,3,4,6-Tetrachlorophenol*

2,3,4,6-Tetrachlorophenol (TCP) is used as an insecticide, sterilizer, wood preserver, and formicide (ant-specific insecticide). The organism that is the most susceptible to its effects is the water flea, and its ECOSAR class is that of phenols. The amount of

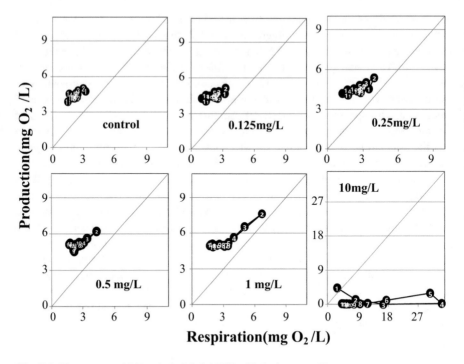

Fig. 7.6 Time course of P/R ratio in 2,3,4,6-TCP-added microcosm N-system

production and respiration decreased in the microcosm with even a 1 mg/L addition, and effects were observed. The influence on the amount of production decreasing to 0 occurred at an addition concentration of 10 mg/L, and it was clear that TCP more strongly influences production than respiration (Fig. 7.6). As a result of holding P and R constant, a test of the significant difference in the P/R ratio was performed using a branching-type ANOVA. Having been subjected to impact assessment, MCP was found, and the m-NOEC was determined to be 1 mg/L. Additionally, as a result of having compared the mesocosm and microcosm test results, it was clear that they were closely correlated.

7.2.2 Mancozeb

Mancozeb was added to the microcosm N-system at concentrations of 0.3 and 3.0 mg/L 16 days after cultivation began (i.e., during the stationary phase). It was determined that there was a statistically significant difference in both the 0.3 and 3 mg/L loads of Mancozeb. However, the lowest effective concentration was determined because it maintained almost stable production and respiration volumes at 0.3 mg/L of added Mancozeb (Fig. 7.7).

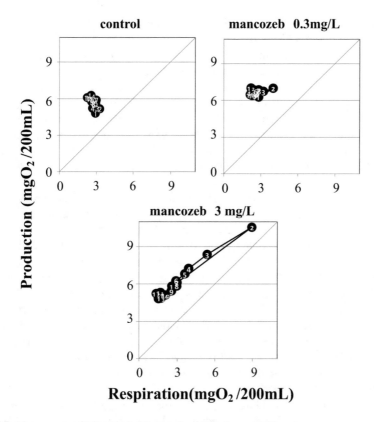

Fig. 7.7 Time course of P/R ratio in Mancozeb-added microcosm N-system

7.2.3 *Thiurum*

Thiurum has been used as a dithiocarbamate-based sterilizer (pesticide) or a repellent against birds. The vulcanized accelerants in the latex are cited as the main use of this chemical other than as a sterilizer. The thiurum concentration added in the micro-cosm tests were set to 0.25, 0.5, 1, 2, 4, and 8 mg/L, and the methanol concentration as the solvent was adjusted to 0.32%. Experiments involving the addition of thiurum to the microcosm were conducted 16 days after the microcosm culture began (i.e., during the stationary phase).

As a result of exposure to thiurum, bacteria, *Lecane* sp., *Philodina erythrophthalma*, and *Aeolosoma hemprichi* tended to multiply, which allowed bacteria to uptake as a nutrient source the methanol from the organic substances added as a solvent, and microanimals predated them. Microalgae, such as *Chlorella* sp., were not affected by the addition of thiurum. Furthermore, the addition of 0.32% of a methanol solvent did not affect the microcosm. *Cyclidium glaucoma* and *Aeolosoma hemprichi* decreased in their populations with the addition of 1.0 mg/L of thiurum. From the single-species culture test, it was made clear that *Aeolosoma hemprichi* was strongly affected by thiurum. However, at the 1.0 mg/L addition concentration, the population of *Aeolosoma hemprichi* decreased before then increasing. The populations of *Philodina erythrophthalma* and *Lecane* sp. increased 2 days after the addition of thiurum. At the thiurum addition concentrations of 4.0 mg/L and 8.0 md/L, the populations of *Philodina erythrophthalma* and *Lecane* sp. first decreased and then increased remarkably. This was because of the monopoly of *Lecane* sp. and *Philodina erythrophthalma* on the food source that occurred due to the extinction of *Cyclidium glaucoma* and *Aeolosoma hemprichi*. Additionally, the increase in the bacterial population after the addition of thiurum was related to the decrease in predation pressure by microanimals, which decreased due to the addition of thiurum. Furthermore, this was because the survival of microanimals recovered in the thiurum-added microcosm. Photolysis and biological degradation were both possible with the addition of thiurum, and it was therefore assumed that microanimals were able to multiply remarkably again.

After analyzing the residual properties of thiurum through high-performance liquid chromatography (HPLC), the thiurum concentration exhibited little change in the control system where there were no microorganisms, and it was shown that photolysis did not occur under the microcosm culture conditions. In the 1.0 mg/L thiurum addition system, the thiurum concentration in supernatant was reduced to half 1 day after addition and decreased to below the detection limit by 2 days later. From this, it was thought that thiurum was biodegraded by the microorganisms in the microcosm within a short period of time after its addition. After analyzing thiurum concentrations in the microorganisms sampled by filtration, the characteristics of thiurum accumulation were not accepted.

From the evaluation of thiurum toxicity in the single-species culture test, the EC_{50} of *Philodina erythrophthalma* and *Aeolosoma hemprichi* were estimated as 6.0 mg/L and 1.5 mg/L, respectively. However, the concentration at which an increased influence was observed in comparison with the control system was determined as the influence concentration in the microcosm test. The concentration at which the influence of thiurum increased in the single-species culture test was also determined to compare both test methods. *Philodina erythrophthalma* was affected at a concentration of 4.0 mg/L of added thiurum in the single-species culture test, the same as in the microcosm test. *Aeolosoma hemprichi* was affected at a concentration of 1.0 mg/L of added thiurum in the single-species culture test, which was also the same as in the microcosm test. Generally, the influence of thiurum was more difficult to observe in the single-species culture test, but it was readily observed in the microcosm test. These findings are thought to have resulted from the differences in the basal media of the microorganism populations between the two methods.

7.3   Herbicides

Selective herbicides control specific weed species while leaving the desired crop relatively unharmed. Meanwhile, nonselective herbicides (sometimes called total weed killers in commercial products) can be used to clear plants from the ground, industrial and construction sites, and railways and railway embankments, as they kill all plant material with which they come into contact. Apart from selective/ nonselective herbicides, other important distinctions among these chemicals include their persistence (i.e., their residual action or how long the product stays in place and remains active), means of uptake (whether it is absorbed by aboveground foliage only, through roots, or by other means), and the mechanism of action. Historically, products such as common salt and other metal salts were used as herbicides; however, these have gradually fallen out of favor, and in some countries, a number of these are banned due to their persistence in soils and toxicity and due to concerns regarding groundwater contamination. Herbicides have also been used in warfare and armed conflicts. Modern herbicides are often synthetic mimics of natural plant hormones, which interfere with the growth of target plants. The term "organic herbicide" has come to mean herbicides intended for organic farming. Some plants also produce their own natural herbicides, such as the genus *Juglans* (walnuts) or the tree of heaven; the actions of natural herbicides and other related chemical interactions are termed allelopathy. Due to herbicide resistance, a major concern in agriculture, a number of products combine herbicides with different means of action. Integrated pest management may use herbicides alongside other pest control methods.

Though herbicides are known to exhibit high toxicity in the aquatic ecosystems that contain biological interactions, material circulation, and energy flow, the environmental risk associated with the use of herbicides remains poorly understood. The microcosm, which is a model microbial ecosystem consisting of a producer, consumers, and decomposers, is useful for evaluating the environmental risk to an ecosystem, and positioning of this model ecosystem as the standard docimasy is thus important.

7.3.1 Linuron

The addition concentration of Linuron was adjusted to 0.01, 0.05, 0.1, 1.0, and 3.0 mg/L, and it was added on the 16th day of the stable state into the microcosm. The endpoints were abundance (structural parameter) and the DO concentration (functional parameter), and the population was measured using an optical microscope and counted from the start of culturing on days 0, 2, 4, 7, 14, 16, 18, 20, 23, and 30, and it was evaluated from the results of B_{16-30} (days 16–30), which was the ratio of abundance and population density (N_{30}) on the 30th day. The DO concentration was measured continuously from the 16th day onward, and the P/R ratio was calculated from the amounts of production (P) and respiration (R).

The m-NOEC of Linuron as a herbicide was determined to be 0.1 mg/L using the functional parameter (Fig. 7.8), but no effect was observed in the time series of biotic succession. It was also determined that the m-NOEC of Linuron as herbicide was in the range of 1–10 mg/L by the structural parameter. The activity of the microcosm increased with the addition of 1 mg/L and decreased with the addition of 10 mg/L of Linuron. The strength of Linuron loading (i.e., the influence concentration) was

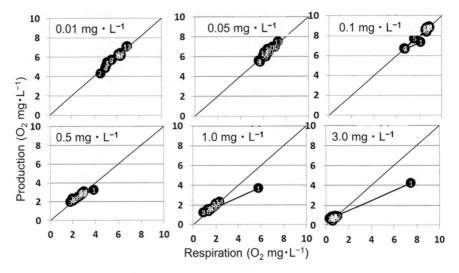

Fig. 7.8 Time course of P/R ratio in Linuron-added microcosm N-system

greater for production than for consumption. Although it is difficult to analyze the functions of production and consumption in the microcosm based on the abundance of each microorganism and measuring of the DO concentration, the above information can be obtained using the integrated Environmental Impact Risk Assessment System (iEIRAS).

7.3.2 Fomesafen

Fomesafen partially inhibits the production of chlorophyll, and it is used for broad leaf weeds, with the obstruction of a photosynthesis mainly serving to damage the cell membrane. Additionally, pesticide registration in Japan is not yet complete. Acetone was used as a solvent and was regulated to concentrations of 0.24% after addition to the TP medium. The addition concentration of Fomesafen was adjusted to 0.5, 0.6, 0.7, 1.0, 2.0, 5.0, 10, 20, 30, and 50 mg/L, and it was added on the 16th day of the stable state into the microcosm. The endpoints were abundance (structural parameter) and DO (functional parameter), and the population was measured using an optical microscope and counted from the start of culturing on days 0, 2, 4, 7, 14, 16, 18, 20, 23, and 30, and it was evaluated from the result of B_{16-30} (days 16–30), which was the ratio of abundance and population density (N_{30}) on the 30th day. The concentration of DO was measured continuously from the 16th day onward, and the P/R ratio was calculated from the amounts of production (P) and respiration (R).

A change in the abundance of microorganisms was observed according to the time series of the structural parameter. All organisms living in all of the experimental systems entered an acute period after the addition of Fomesafen. The abundance of *Lecane* sp. decreased at addition concentrations of more than 20 mg/L of fomesafen, the number of *Aeolosoma hemprichi* decreased significantly at concentrations of more than 0.6 mg/L, *Cyclidium glaucoma* perished, and the population of *Philodina erythrophthalma* increased and then decreased from herbicide addition in the sub-acute period. The abundance of *Lecane* sp. decreased at addition concentrations of 5 mg/L or more during the chronic period, *Aeolosoma hemprichi* significantly decreased at concentrations of more than 0.6 mg/L, *Cyclidium glaucoma* perished at concentrations of more than 2 mg/L, and *Philodina erythrophthalma* increased and decreased in abundance through the addition of fomesafen. As for the phytoplankton, *Chlorella* sp. decreased to about half of its original population size in all microcosms, and there were few changes. The population of *Scenedesmus quadricauda* tended to increase incrementally, but it exhibited chronicity even when there was a tendency toward increasing and then decreasing at concentrations of more than 30 mg/L. *Tolypothrix* sp. increased in abundance but decreased with the addition of more than 20 mg/L of Fomesafen.

From the viewpoint of B_{16-30} (days 16–30 of the experiment), there was no significant decrease in the abundance of any organisms at an addition concentration of 0.5 mg/L, but the population of zooplankton decreased by more than half with concentrations of greater than 0.6 mg/L. From this, it was estimated that the m-NOEC was 0.5 mg/L for the structural parameter. Additionally, a strong influence on zooplankton was observed with the addition of Fomesafen, and a decrease in the abundance of *Chlorella* sp. was observed, while an increase was recognized for the populations of both *Scenedesmus quadricauda* and *Tolypothrix* sp. This depended on the difference in the resistance of the microbial species in the microcosm, and it is thought that the presence or absence of the influence differed and might be produced by different choices of herbicide, resulting in different effects on a given plant.

There was no observed difference in the functional parameter (DO concentration) at addition concentrations of 0.6 mg/L or 10 mg/L, but activity decreased at 30 mg/L, and the DO concentration decreased with the addition of 50 mg/L of fomesafen (Fig. 7.9). The P/R ratio was stable at ~1 for addition concentrations of 0.6 mg/L and 10 mg/L, but some variation was observed at a concentration of 30 mg/L, and the value greatly declined with the addition of 50 mg/L of fomesafen (Fig. 7.9). From these results, it can be estimated that the m-NOEC was approximately 30 mg/L for the functional parameter.

7.3.3   Simazine

Simazine only affected the growth of *Tolypothrix* sp., which was measured for 15 days after the addition of the herbicide, and it decreased to 1/40th of its original value with the addition of 0.08 mg/L or more of Simazine. The influence of this chemical was mitigated by the diversity of the constituent species in the microcosm. Simazine remained very stable in the microcosm. It was found that the toxicity—the

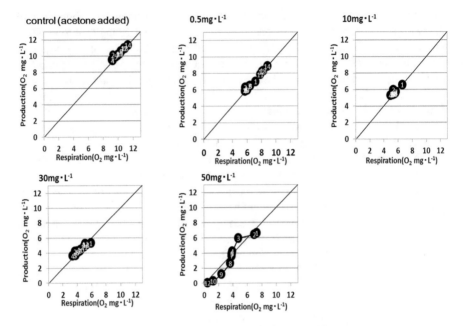

Fig. 7.9 Time course of P/R ratio in Fomesafen-added microcosm N-system

influence and stability of pesticides in the aquatic model ecosystem—could be evaluated from the behavior of the constituent microorganisms and pesticides. Addition concentrations of simazine were set to 0.08, 0.16, 0.32, and 0.64 mg/L. In the control system, only methanol was added to the system, and both methanol-addition and no-addition systems were constructed to compare the influence of methanol as a pesticide solvent.

Using the structural parameter, there was no observed effect on the population of the methanol-addition system in comparison with the no-addition system, which was used as a control. There was no effect on the populations of constituent microorganism with the addition of 0.16, 0.32, or 0.64 mg/L of Simazine, except for a decrease in the bacterial population 2 days after the addition of 0.64 mg/L of simazine. Changes were not observed in either the amount of respiration (R) or in the amount of production (P) in the control system (methanol no-addition), but, in the control system with methanol added, both the amounts of respiration and production increased. In addition systems with low simazine concentrations, changes similar to those observed in the methanol-addition control system were recognized, but a decrease in the amount of production was observed as the concentration of simazine increased. Although there was no change in the P/R ratio, which ranged between 1.6 and 1.8, in either of the control systems (i.e., both with and without methanol), the P/R ratio decreased to 0.8 after the addition of 0.64 mg/L of Simazine (Fig. 7.10).

In the multivariation analysis of the amounts of respiration and production after the addition of simazine, a meaningful decrease in the amount of production was recognized between the 0.64 mg/L simazine load and the control with added

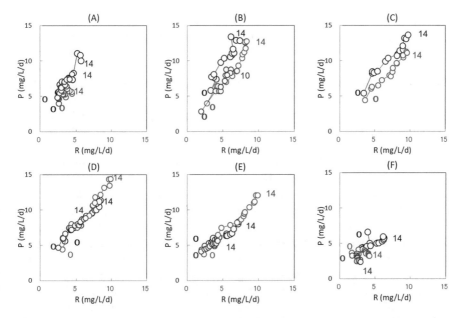

Fig. 7.10 Time course of P/R ratio in Simazine-added microcosm N-system. (**a**) Control. (**b**) Control + methanol. (**c**) 0.08 mg/L. (**d**) 0.16 mg/L. (**e**) 0.32 mg/L. (**e**) 0.64 mg/L

methanol from the 4th day after addition onward, and both the low Simazine load (0.64 mg/L) and high load (16 mg/L) systems were defined from the 1st day after the addition of the herbicide. In these cases, the amount of production did not decrease meaningfully, whereas the P/R ratio decreased in the high load system. The toxicity of Simazine was not recognized as having an influence on the populations of microorganisms at the concentration ranges employed in this experiment. Thus, the amount of production appears to have significantly decreased as a direct effect of simazine alone, and there was no influence on the populations of microorganisms, and the P/R ratio tended to decrease. It is necessary to use the measured amounts of production and respiration together to investigate more than simply the changes in the populations of microorganisms in an ecosystem when performing an ecosystem impact assessment of pesticides because there are differences in ecosystem responses depending upon the kind of pesticide used and the methods of assessment.

7.3.4   Benthiocarb

Benthiocarb affected the growth of *Cyclidium glaucoma* from a concentration of 1.0 mg/L in the two-species cultivation test and from a concentration of 2.0 mg/L in the microcosm test. The influence was mitigated by the diversity of constituent species in the microcosm. *Aeolosoma hemprichi* was eliminated, and the number of *Chlorella* sp. decreased to one-third its original value at Benthiocarb concentrations above 0.5 mg/L. In the microcosm, the concentration of benthiocarb decreased to 30% for 15 days after the addition of 1.0 mg/L. It was found that the toxicity of Benthiocarb could be evaluated based on the behaviors of the constituent microorganisms and pesticides. Addition concentrations of benthiocarb were set to 1.0, 2.0, 4.0, and 8.0 mg/L. As the controls, only methanol-addition and methanol no-addition systems were constructed to compare the influence of methanol as a pesticide solvent.

Using the structural parameter, there was no difference observed in the populations of microorganisms between the methanol-addition and no-addition control systems. In the microcosms where 1.0–4.0 mg/L of Benthiocarb was added, there was also no significant influence on the population; however, both *Aeolosoma hemprichi* and *Philodina erythrophthalma* perished with the addition of 8.0 mg/L of Benthiocarb. The functional parameter and the amounts of respiration (R) and production (P) increased with the addition of methanol, but not only was there no observed influence on the ecosystems of the methanol-addition system and in the case of Benthiocarb loading, but there was also no difference recognized at the high load of 8.0 mg/L of benthiocarb. The P/R ratio did not change with the addition of Benthiocarb (Fig. 7.11). Differences in the amounts of production and respiration on each measurement day after the addition of Benthiocarb were not recognized as statistically significant in comparison with either of the methanol control systems according to the results of ANOVA. No significant difference in the P/R ratio was recognized between the methanol-addition system and Benthiocarb-load systems, either. With respect to the toxicity of Benthiocarb to animals, it was reported that the LD_{50} values for carp and *Daphnia* are 3.6 mg/L and 1.7 mg/L, respectively, but the concentration responsible for the mortality of microanimals in the microcosm was approximately 8.0 mg/L. However, no influence from Benthiocarb itself was observed on the amounts of production or respiration at the concentration in which microanimal populations became extinct.

7.3.5   Propyzamide

Propyzamide is a benzamide herbicide used for the weeding of a particularly gramineous plant and works by obstructing the microtubular composition that is necessary for cell division. Acetone was used as a solvent and regulated to a concentration of 0.24% after addition to the TP medium. The addition concentration of Propyzamide was adjusted to 0.1, 0.2, 1, 3, and 5 mg/L, and it was added on the 16th day of the stable state into the microcosm. The endpoints were abundance (structural parameter) and the concentration of DO (functional parameter), and the population was measured using an optical microscope and counted from the start of culturing on days 0, 2, 4, 7, 14, 16, 18, 20, 23, and 30, and it was evaluated from the results of B_{16-30} (days 16–30), which was the ratio of abundance and population

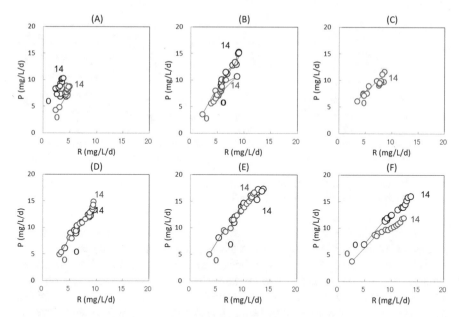

Fig. 7.11 Time course of P/R ratio in Benthiocarb-added microcosm N-system. (**a**) Control. (**b**) Control + methanol. (**c**) 1.0 mg/L. (**d**) 2.0 mg/L. (**e**) 4.0 mg/L. (**f**) 8.0 mg/L

density (N_{30}) on the 30th day. The DO concentration was measured continuously from the 16th day, and the P/R ratio was calculated from the amounts of production (P) and respiration (R).

Based on examination of the population density (N_{30}) in each herbicide addition system, no extinction of zooplankton was observed, but a tendency to decrease was recognized for all microbiota at a Propyzamide concentration of 5 mg/L. Based on the microorganismic abundance (B_{16-30}) at each concentration, *Cyclidium glaucoma* and *Lecane* sp. decreased at high addition concentrations. This is because bacteria, the prey of both *Cyclidium glaucoma* and *Lecane* sp., decreased in abundance with the addition of Propyzamide. Additionally, the phytoplankton were hardly affected, except for *Scenedesmus quadricauda*. The selectivity of Propyzamide is that it works on only rice seeds, but a future examination of the differing influence of this chemical on phytoplankton in the microcosm is necessary.

Significant changes in the concentration of DO were not observed with the addition of 1 mg/L of Propyzamide, but production and respiration amounts increased at addition concentrations of more than 3 mg/L (Fig. 7.12). Additionally, it may be said that the activity *per* cell of both microanimals and microalgae increased at concentrations of more than 1 mg/L because the abundance decreased. However, the P/R ratio was stable, so it was estimated that there was no effect on the microcosm ecosystem. Furthermore, the NOEC value is greater than 2.2 mg/L with respect to reproduction (NOECREP) for 21 days in *Daphnia magna* and 0.32 mg/L for 3 days in the chlorophycean, *Pseudokirchneriella subcapitata*. It is thought that

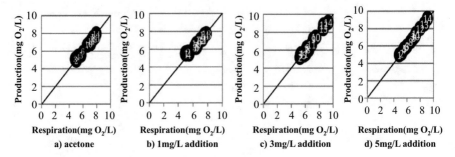

Fig. 7.12 Time course of P/R ratio in Propyzamide-added microcosm N-system. (**a**) Acetone. (**b**) 1 mg/L addition. (**c**) 3 mg/L addition. (**d**) 5 mg/L addition

the influence on the ecosystem was reduced because the interaction among the organisms in the microcosm functioned where one ecosystem is constructed.

From these outcomes, it was estimated that the m-NOEC of the structural parameter was greater than 5 mg/L because, while mainly the zooplankton were affected, extinction was not observed. Similarly, the m-NOEC of the functional parameter was also estimated as being greater than 5 mg/L, and there was an observed tendency toward increased production and respiration in all addition systems. Populations of phytoplankton and zooplankton decreased with the increasing addition concentration of propyzamide, but the activity *per* individual rose simultaneously, and it was thought that equilibrium was maintained. Influence on ecosystem was relaxed than a single creature by the microbial interaction in microcosm, and it was made clear that a difference was observed in comparison with the effects of Propyzamide in the single-species test.

7.3.6 *2-Methyl-4-Chlorophenoxyacetic Acid (MCPA)*

2-Methyl-4-chlorophenoxyacetic acid (MCPA), a chemical substance similar to auxin, disturbs hormonal balance and is therefore a hormonal drug. The addition concentration of MCPA was adjusted to 0.1, 0.2, 0.3, 0.4, and 0.5 mg/L, and it was added on the 16th day of the stable state into the microcosm. The endpoints were population density (structural parameter) and the concentration of DO (functional parameter), and the population was measured using an optical microscope and counted from the start of culturing on days 0, 2, 4, 7, 14, 16, 18, 20, 23, and 30, and it was evaluated from the results of B_{16-30} (days 16–30), which was the ratio of abundance and population density (N_{30}) on the 30th day. The DO concentration was measured continuously from the 16th day, and the P/R ratio was calculated from the amounts of production (P) and respiration (R).

Extinction of the constituent microorganism was not confirmed from the evaluation of the population density, nor was ecological collapse observed in the

Fig. 7.13 Time course of P/R ratio in MCPA-added microcosm N-system. (**a**) 0.3 mg/L addition. (**b**) 0.4 mg/L addition

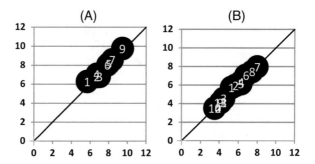

microcosm. The populations of *Aeolosoma hemprichi* and *Cyclidium glaucoma* decreased remarkably by the 30th day. The influence of MCPA was recognized as a tendency to decrease population sizes at concentrations ranging from 0.3 to 0.4 mg/L. Most species were observed to increase in population size with the addition of 0.2 mg/L of MCPA because a system is opposed to this for weed killer MCPA, and it may be said that populations increased because they were active. Therefore, the m-NOEC of the structural parameter was estimated to be 0.4 mg/L. There was no observed difference in the functional parameters of the control and addition systems until an addition concentration of 0.3 mg/L was reached. However, activity gradually decreased at an addition concentration of 0.4 mg/L, and the results were similar at concentrations of more than 0.5 mg/L. While the P/R ratio of the control system was ~1 and stable, the activity tended to increase at a concentration of 0.3 mg/L and decrease at a concentration of 0.4 mg/L loading (Fig. 7.13). Thus, it can be estimated that the m-NOEC of the functional parameter is 0.4 mg/L.

7.3.7   Alachlor

Alachlor is an acid amide herbicide, and it is considered to wither plants by interrupting normal cell division at the position of new growth by synthesizing inhibitors of very-long-chain fatty acids. On the 16th day after the start of the microcosm culture, Alachlor was adjusted to concentrations of 0.1, 1, and 4 mg/L and added. Evaluation of production (P) and respiration (R) in the microcosm showed that the m-NOEC of Alachlor was present in the range of 1–4 mg/L (Fig. 7.14).

7.3.8   3-(3,4-Dichlorophenyl)-1,1-Dimethylurea (DCMU)

3-(3,4-Dichlorophenyl)-1,1-dimethylurea (DCMU) is a typical, urea-based, photosynthesis-inhibiting herbicide. Its herbicidal action is mild, and it is used as a

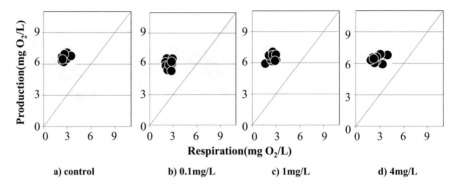

Fig. 7.14 Time course of P/R ratio in Alachlor-added microcosm N-system. (**a**) Control. (**b**) 0.1 mg/L. (**c**) 1 mg/L. (**d**) 4 mg/L

common soil conditioner. The m-NOEC in the microcosm test was evaluated to be 0.05 mg/L using the structural parameter and to be 0.1 mg/L based on the functional parameter.

7.3.9 Pentachlorophenol

The effect of pentachlorophenol (PCP) on microcosms was investigated by Dr. Sugiura. PCP was added at the start of the microcosm culture (day 0), and the abundance of organisms and amounts of respiration and production were tracked. With the addition of PCP, both the respiration volume and production volume became larger than in the control system. The m-NOEC of PCP was estimated to be 0.07 mg/L (Sugiura 1992).

7.3.10 Atrazine

The effect of atrazine was investigated by Dr. Sugiura. Atrazine was added and adjusted to concentrations of 0.07, 0.1, and 0.15 mg/L at the start of the microcosm culture (day 0). The locus of DO differed depending upon the amount of atrazine added. With the addition of atrazine, both the respiration volume and production volume became larger than in the control system. The m-NOEC of atrazine was estimated to be 0.03 mg/L (Sugiura 1992).

7.3.11 3,4-Dichloroaniline

The effect of 3,4-dichloroaniline (3,4-DCA) was investigated by Dr. Sugiura. 3,4-Dichloroaniline was added at the start of the microcosm culture (day 0), and the number of organisms (abundance) and amounts of respiration and production were tracked. With the addition of 3,4-DCA, both the respiration volume and production volume became larger than in the control system. The m-NOEC of 3,4-DCA was estimated to be 1 mg/L (Sugiura 1992).

7.3.12 Paraquat

The effect of paraquat was investigated by Dr. Sugiura. Paraquat was added at the start of the microcosm culture (day 0), and the number of organisms and amount of respiration and production were tracked. The m-NOEC was estimated to be 0.01 mg/L (Sugiura 1992).

7.3.13 2,4,5-Trichlorophenoxyacetic Acid (2,4,5-T)

The effect of 2,4,5-trichlorophenoxyacetic acid (2,4,5-T) was investigated by Dr. Sugiura. 2,4,5- Trichlorophenoxyacetic acid was added at concentrations of 0.1, 1.0, 10, and 100 mg/L at the start of the microcosm culture (day 0). The change over time in the population density and community metabolism with the addition of 10 mg/L did not differ from those of the control system. In the 100 mg/L addition system, *Philodina erythrophthalma*, *Aeolosoma hemprichi*, *Scenedesmus quadricauda*, and *Tolypothrix* sp. perished, and the change over time in the community metabolism also differed greatly from the control system. From these results, the addition of 10–100 mg/L of 2,4,5-T was estimated to have affected this system (Sugiura 1992).

7.3.14 Cafenstrole

On the 16th day after the start of the microcosm culture (stationary phase), cafenstrole was added at concentrations of 0, 0.5, 1.0, 2.5, and 5.0 mg/L. The influence exerted by this chemical on the microorganisms constituting the microcosm depended upon the concentration added. At an addition concentration of 0.5 mg/L, *Cyclidium glaucoma*, *Lecane* sp., and *Aeolosoma hemprichi* began to be affected, with *Aeolosoma hemprichi* being affected at 1.0 mg/L, and both *Aeolosoma hemprichi* and *Cyclidium glaucoma* disappeared within 2 days after the addition of

2.5 mg/L of cafenstrole. With the addition of 5.0 mg/L of cafenstrole, *Aeolosoma hemprichi* disappeared within 2 days after addition, and both *Lecane* sp. and *Cyclidium glaucoma* disappeared from the system on the 7th day following chemical addition. In this way, predatory microanimals that contribute to the material circulation and energy flow in the microcosm were deemed susceptible to cafenstrole.

7.4   Pesticides

Pesticide has been widely used and it improved crop yields dramatically, and our life has become rich. On the other hand, pesticide is recognized that it influences lives including humans very seriously such as defoliant which has been sprayed in the Vietnam War. Biological toxicity of pesticide is regulated by various bioassay tests, but its ecosystem toxicity is not enough cleared.

7.4.1   Chlorpyrifos

Dursban 4E, an emulsification pesticide, was prepared as Chlorpyrifos. This is because Leeuwangh et al. (1994) and Daam and van den Brink (2007) used Dursban 4E, which is an emulsification pesticide product as Chlorpyrifos, Daam et al. (2008) used Dursban 40EC, and López-Mancisidor et al. (2008a, b) used Chas 48EC, an emulsification pesticide product, in their mesocosm experiments. Chlorpyrifos was added at the start of microcosm cultivation, and the microcosm test was conducted with concentrations of 125, 250, 500, and 1000 µg/L, with the common ratio set to 2.

At any concentration loaded, the P/R ratio was around 1, but the production and respiration amounts were significantly lower at a Chlorpyrifos loading of 1000 µg/L (Fig. 7.15). The NOEC of Chlorpyrifos was estimated to be 500 µg/L. Even if an organism perished with the addition of 10,000 µg/L of Chlorpyrifos, there was no observed impact on the production or respiration at all. Chlorpyrifos is a neurotoxin and reduces the activity of animals and causes a drop in the respiration amount *per* individual. It was thought that the amount of respiration in the entire microanimal population in the microcosm decreased because the population decreased with a rise in death rates. However, it was thought that no influence appeared in the amounts of production and respiration when chlorpyrifos was added during the stationary period because the respiration amount had been supplemented by decomposition of the microanimals and the increase in the abundance of many organisms with the decrease in predation. On the other hand, the community structure of the system changed when chlorpyrifos was added when the culture began because Chlorpyrifos drove the succession of microanimals in the microcosm, and it is thought that, as a result, an influence on production and respiration appeared.

The NOEC of Chlorpyrifos in the mesocosm test was reported as 0.06 µg/L. However, NOEC of the microcosm test was 500 µg/L, which is much higher.

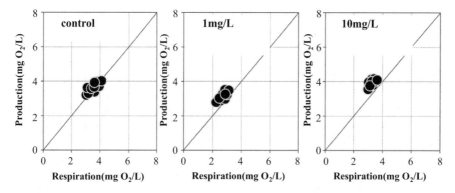

Fig. 7.15 Time course of P/R ratio in Chlorpyrifos-added microcosm N-system

Hydrolysis is high in Chlorpyrifos, and the quality of soil adsorption characteristics is also high, and it is thought that it is stabilized while adsorbed in the soil. Therefore, it is thought that Chlorpyrifos remains behind within a system for a long term by soil adsorption and stabilization. Chlorpyrifos is adsorbed to the bottom sediment when Chlorpyrifos density decreases by hydrolysis detaching by chemical equilibrium, and even one revelation being supplied to the mesocosm test where there is bottom sediment underwater. Because there was no soil in the microcosm, the residual property of Chlorpyrifos was low, and it was thought that it was less toxic than in the mesocosm test in which there was soil. It is necessary to examine the analytical methods that noted the quality of soil and bottom sediment adsorption characteristics in the future.

7.4.2   Fenitrothion

Fenitrothion was adjusted so that acetone levels became 0.05% after addition to the culture fluid in diluted form. The concentrations of Fenitrothion were set to 0.01, 0.10, 1.00, and 10.00 mg/L. The Fenitrothion was added 16 days after microcosm culturing began. As a control, an acetone-only addition system and a no-addition system were constructed to compare the influence of acetone as a solvent for pesticides.

Based on the structural parameter, there was no influence of acetone itself (i.e., in the acetone-only system) on the populations of microorganisms, in comparison with the no-addition system, both of which served as control systems. There was no remarkable change observed in the populations of each species, in comparison with these control systems, at concentrations of 0.01 mg/L of Fenitrothion, but the population size of *Aeolosoma hemprichi* decreased with the addition of 0.10 mg/L and 1.00 mg/L. There was no observed influence on the other microorganisms at these concentrations of Fenitrothion. However, *Aeolosoma hemprichi*, *Philodina erythrophthalma*, and *Lecane* sp. perished, and the population of *Cyclidium glaucoma* decreased remarkably and then recovered with the addition of 10 mg/L of fenitrothion. Thus, there was a recognized influence from this pesticide on microanimals in the microcosm. In contrast, the microalgae, *Chlorella* sp., *Scenedesmus quadricauda*, and *Tolypothrix* sp., were not influenced by the addition of less than 10 mg/L of Fenitrothion. The functional parameter, the quantity of production (P) and respiration (R), increased in comparison with the no-addition control system. There was no difference between the P/R ratios of systems with Fenitrothion concentrations ranging from 0.01 to 10 mg/L and the acetone-only control system. The P/R ratio did not change with the addition of Fenitrothion (Fig. 7.16).

To understand the effects of Fenitrothion, an ANOVA was performed between the acetone-only control system and each Fenitrothion addition system, but no meaningful difference was recognized in the amounts of production and respiration or in the P/R ratios of each microcosm; the influence of the Fenitrothion concentration remained the same. The NOEC of Fenitrothion varied widely by species; it is reported as 0.009 mg/L for *Daphnia magna*, a very small amount, whereas it was ~1.0 mg/L for rotifers and microalgae. To focus on the constituent animals of the microcosm, because the abundance of *Aeolosoma hemprichi* decreased with the addition of 0.1 mg/L of Fenitrothion and *Aeolosoma hemprichi* and two species of rotifer, *Philodina erythrophthalma* and *Lecane* sp., perished with the addition of 10 mg/L of Fenitrothion, it may be said that influence at the same level as the aquatic animal that tolerance is relatively strong. However, even at the concentration at which constituent animals in the microcosm perished, there was no observed influence by Fenitrothion alone on the amounts of production and respiration.

7.4.3   Lindane

Lindane was added to the microcosm N-system on the 16th day after cultivation began (stationary phase) at concentrations of 0.003, 0.03, and 0.12 mg/L. There was no statistically significant difference between the addition and control systems, and stable production and respiration amounts were maintained that were equal to those in the control system, at Lindane concentrations of 0.003, 0.03, and 0.12 mg/L (Fig. 7.17). It was estimated that the m-NOEC of Lindane is greater than 0.12 mg/L.

7.4.4   Carbendazim

A benrate hydration agent was supplied for investigation. A benrate hydration agent is a main product of Carbendazim, composed primarily of approximately 50% of benomyl, and the benomyl is converted into Carbendazim underwater, having a half-life of 2 h. The addition concentration of Carbendazim was adjusted to 0.1, 1.0, 3, 5, and 10 mg/L, and it was added on the 16th day of the stable state into the microcosm.

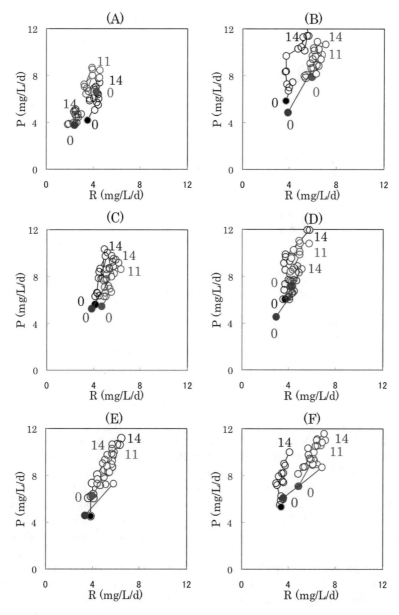

Fig. 7.16 Time course of P/R ratio in Fenitrothion-added microcosm N-system. (**a**) Control. (**b**) Control + aceton. (**c**) 0.01 mg/L. (**d**) 0.10 mg/L. (**e**) 1.00 mg/L. (**f**) 10.00 mg/L

The endpoints were abundance (structural parameter) and the concentration of DO (functional parameter), and the population was measured using an optical microscope and counted from the start of culturing on days 0, 2, 4, 7, 14, 16, 18, 20, 23, and 30, and it was evaluated from the results of B_{16-30} (days 16–30), which was

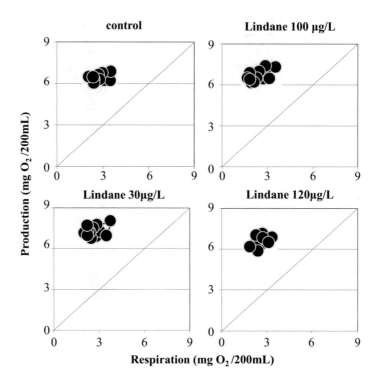

Fig. 7.17 Time course of P/R ratio in Lindane-added microcosm N-system

the ratio of abundance and population density (N_{30}) on the 30th day. The DO concentration was measured continuously from the 16th day, and the P/R ratio was calculated from the amounts of production (P) and respiration (R).

The structural parameter was evaluated according to the categories of acute influence B_{16-20} (0–4 days), subacute influence B_{20-23} (5–7 days), chronic influence B_{23-30} (8–14 days), and abundance of microorganisms (N_{30}) on the 30th day after culturing began. As a result, no large decreases or extinctions of species were seen in the periods of subacute or acute influence, but a large decrease in the abundance of *Cyclidium glaucoma* and *Aeolosoma hemprichi* was observed in the addition system with more than 1 mg/L of Carbendazim during the period of chronic influence. Additionally, phytoplankton did not exhibit any remarkable increase or decrease, and no influence upon them was recognized. Because Carbendazim is a sterilizer, at first this influenced the bacteria which act as decomposers in the microcosm, and it influenced zooplankton predation on them afterward. It is thought that there was little impact on the phytoplankton producers. Decreases in the abundance and extinction of zooplankton were observed for N_{30} as concentrations increased, but an increase in the abundance of *Philodina erythrophthalma* was also recognized in the 10 mg/L addition system. This is considered to be due to the extinction of

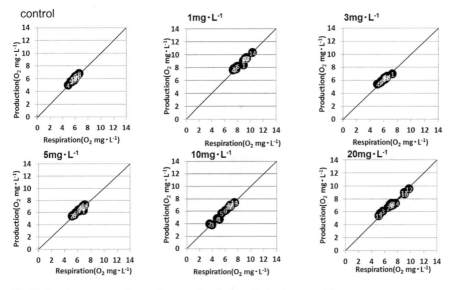

Fig. 7.18 Time course of P/R ratio in Carbendazim-added microcosm N-system

Cyclidium glaucoma and *Aeolosoma hemprichi*, and *Philodina erythrophthalma* supplemented the consumers and increased to maintain the system. As a functional parameter, the amplitude of the DO damping as the concentration increased, along with the activity of the system, indicated a tendency to decline slightly. However, all systems were stable, with P/R ratios of ~1. The m-NOEC of Carbendazim, as a fungicide, was detected within the range of 0.1–1 mg/L in the structural parameter and as more than 10 mg/L in the functional parameter (Fig. 7.18).

7.4.5   Dimethoate

Dimethoate is organophosphorus pesticide that inhibits a nerve transduction system (acetylcholinesterase (AChE) activity) necessary for transmitting information for the life support of the pest and is mainly used for exterminating pests of fruit trees. The addition concentration of Dimethoate was adjusted to 0.05, 0.1, 0.2, 0.5, 1.0, and 2.0 mg/L, and it was added on the 16th day of the stable state into the microcosm. The endpoints were abundance (structural parameter) and the concentration of DO (functional parameter), and the population was measured using an optical microscope and counted from the start of culturing on days 0, 2, 4, 7, 14, 16, 18, 20, 23, and 30; it was evaluated from the results of B_{16-30} (days 16–30), which was the ratio of abundance and population density (N_{30}) on the 30th day. The DO concentration was measured continuously from the 16th day, and the P/R ratio was calculated from the amounts of production (P) and respiration (R).

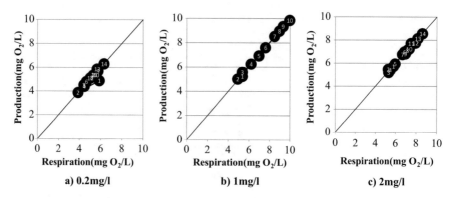

Fig. 7.19 Time course of P/R ratio in Dimethoate-added microcosm N-system. (**a**) 0.2 mg/L. (**b**) 1 mg/L. (**c**) 2 mg/L

The population (structural parameter) of phytoplankton decreased over time after the addition of dimethoate, and only *Tolypothrix* sp. maintained its population at all addition concentrations. The abundance of *Cyclidium glaucoma* decreased sharply at concentrations of 0.5 mg/L or more during the subacute period, and extinction was observed at a concentration of 1 mg/L during the acute period. A remarkable difference was observed in the abundance of phytoplankton and zooplankton from B_{16-30} (days 16–30). As for the phytoplankton, *Scenedesmus quadricauda* and *Tolypothrix* sp. tended to increase in abundance, relative to the control system, at concentrations of more than 1 mg/L, but *Chlorella* sp. did not. All species of zooplankton decreased in abundance, as compared with the control system, at concentrations of more than 0.2 mg/L. It is thought that differences might occur among the influences on species due to the mechanisms of the insecticide used for pest extermination. As a functional parameter, a difference in the DO concentration was not recognized between dimethoate addition systems with concentrations between 0.05 mg/L and 0.2 mg/L and the control system, but the activity rose with 0.5 mg/L of added dimethoate. Additionally, activity became uniform at concentrations ranging from 0.5 mg/L to 1 mg/L, and an active decline was observed at a concentration of 2 mg/L. However, it is necessary to examine systems with concentrations of more than 2 mg/L because there is not a system to decrease than control system on DO level. As for P/R ratio, activity also rose using test additions ranging from 0.2 mg/L to 1.0 mg/L of dimethoate, but it began to decrease when more than 1.0 mg/L and up to 2.0 mg/L was used (Fig. 7.19).

The m-NOEC of the structural parameter was estimated to be less than 1 mg/L because of the extinction of *Cyclidium glaucoma*, and m-NOEC of the functional parameter was in the range of 1 mg/L to 2 mg/L. Because the abundance of phytoplankton and the amount of production decreased over time, it is thought that there is no influence on the function *per* individual phytoplankton. On the other hand, it is thought that the function *per* individual decreased in zooplankton because one species perished and other species maintained their populations with decreased consumption.

7.4.6   Dinotefuran

In recent years, the neonicotinoid pesticide, Dinotefuran, has been widely used as one of the essential pesticides, but information regarding its influence on the ecosystem hydrosphere through drainage and river water remains lacking. Here, an environmental impact assessment of Dinotefuran, a neonicotinoid pesticide, on microbial community function and structure, was conducted using the experimental flask-sized microcosm. To assess its influence, Dinotefuran was added at several concentrations on the 16th day after microcosm cultivation began. Plankton observation using an optical microscope was conducted on days 0, 2, 4, 7, 14, 16, 18, 20, 23, and 30 after cultivation began, and the change observed in the populations in the flask at each addition concentration was used as the structural parameter. Additionally, the DO concentration was measured consecutively from the 16th day of the culture period onward to evaluate ecosystem function, using the P/R ratio as the functional parameter.

The results based on the structural parameter revealed that the extinction of species was not observed in all systems. A species less than 0.5 for B_{16-30} (the population ratio in the period ranging from the 16th day to the 30th day of the experiment) was recognized at Dinotefuran concentrations of 0.25 and 1.0 mg/L. A decrease in the population of zooplankton was observed in all addition systems from N_{30} (the abundance on the 30th day). Additionally, the population ratio of all zooplankton was less than 0.5 at 1.0 mg/L of added Dinotefuran. For the functional parameter, it was judged that there was no influence in all addition systems based on statistical analyses (Fig. 7.20). The P/R ratios of all systems converged upon a similar value as that of the control system. From these results, the m-NOEC of Dinotefuran was assessed to be over 1.0 mg/L.

7.4.7   4-Chlorophenol

The effect of 4-Chlorophenol was investigated by Dr. Sugiura. 4-Chlorophenol was added at the start of microcosm culturing (day 0), and the number of organisms and amounts of respiration and production were continuously tracked. The m-NOEC of 4-Chlorophenol was estimated to be 0.15 mg/L (Sugiura 1992).

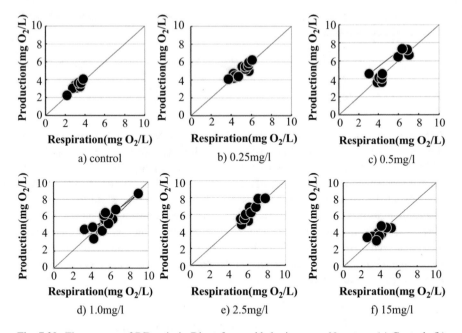

Fig. 7.20 Time course of P/R ratio in Dinotefuran-added microcosm N-system. (**a**) Control. (**b**) 0.25 mg/L. (**c**) 0.5 mg/L. (**d**) 1.0 mg/L. (**e**) 2.5 mg/L. (**f**) 15 mg/L

7.4.8 Methoxychlor

The effect of Methoxychlor was investigated by Dr. Sugiura. Methoxychlor was added at the start of the microcosm culture (day 0), and the number of organisms and amounts of respiration and production were continuously tracked. The m-NOEC of Methoxychlor was estimated to be 0.2 mg/L (Sugiura 1992).

7.4.9 β-Hexachlorocyclohexane (β-HCH)

The effect of β-Hexachlorocyclohexane (β-HCH) was investigated by Dr. Sugiura. β-Hexachlorocyclohexane was added at the start of the microcosm culture (day 0) and was adjusted to 0.1, 0.3, 1.0, and 3.0 mg/L. When 0.1 mg/L of β-HCH was added, the populations of *Cyclidium glaucoma*, *Philodina erythrophthalma*, and *Lecane* sp. increased during the early transition, when compared to the control system. Similarly, in the 1.0 mg/L addition system, the numbers of bacteria and *Cyclidium glaucoma* were higher than those of the control system. Conversely, in the 3.0 mg/L addition system, the populations of *Philodina erythrophthalma*, *Lecane* sp., and *Chlorella* sp. were lower than those of the control system. However, as the transition progressed, each species showed a tendency to approach the population size of the control system at each concentration, and the extinction of organisms did not occur within the concentration range of the microcosm test. The time series of community metabolism also exhibited nearly the same trajectory as in the control system, and the ratio of respiration volume to production volume (P/R ratio) after the 10th day from the start of the culture also approached 1. From these results, β-HCH was not estimated to affect this system in the concentration range examined (Sugiura 1992).

7.4.10   γ-Hexachlorocyclohexane (γ-HCH)

The effect of γ-Hexachlorocyclohexane (γ-HCH) was investigated by Dr. Sugiura. At the start of the microcosm culture (day 0), γ-HCH was adjusted to 0.01, 0.1, 0.5, 1.0, and 5.0 mg/L and added to the system. As in the case of β-HCH, the numbers of *Cyclidium glaucoma*, *Philodina erythrophthalma*, and *Lecane* sp. were higher in the early stage of transition at concentrations of 0.01, 0.1, and 0.5 mg/L when compared with the control system. However, as the transition progressed, the control system population was approached in all of the addition systems. The change over time in community metabolism also exhibited almost the same trajectory as in the control system. From this result, it was assumed that γ-HCH does not affect this system in the concentration range examined (Sugiura 1992).

7.4.11   Dichlorodiphenyltrichloroethane (DDT)

The effect of Dichlorodiphenyltrichloroethane (DDT) was investigated by Dr. Sugiura. At the start of the microcosm culture (day 0), DDT was adjusted to 0.01, 0.05, 0.1, and 0.5 mg/L and added to the system. The addition of DDT did not affect the population density of the constituent organisms, nor did it appear as a change in the time series of the community metabolism relative to the control system until the addition of 0.5 mg/L (Sugiura 1992).

7.4.12   λ-Cyhalothrin

λ-Cyhalothrin is a synthetic pyrethroid insecticide. It acts on the central and peripheral nervous systems and acts to inhibit neurotransmission. On the 16th day after the start of the microcosm culture, λ-Cyhalothrin was adjusted to 1 ng/L and 10 ng/L and added to the system. As a result of performing a one-way ANOVA on the production and respiration, which are functional parameters, it was determined that there was no statistical difference between the control system and λ-Cyhalothrin addition system at any concentration (Sugiura 1992).

7.5   Endocrine-Disrupting Chemicals

Endocrine systems are found in all vertebrate and most invertebrate species. The endocrine system is made up of glands, which secrete hormones to body fluids, and receptor cells, which detect and react to the hormones. The hormones act as chemical messengers. Hormones bind to cells that contain matching receptors in or on their surfaces, much like a key would fit into a lock. Disruption of the endocrine system can occur in various ways. Some chemicals mimic a natural hormone, fooling the body into overresponding to the stimulus or responding at inappropriate times. Other endocrine disruptors block the receptor site on a cell. Still others directly stimulate or inhibit the endocrine system and cause overproduction or underproduction of hormones. The substances that exhibit these effects are known as endocrine-disrupting chemicals (EDCs) or hormone-disrupting chemicals (HDCs). The most conspicuous EDCs are those that affect reproduction. Endocrine-disrupting chemicals have been demonstrated to markedly affect animal populations in coastal environments.

Hormone-disrupting chemicals, or environmental endocrine-disrupters, as they are also known, are the materials that are doubted to disturb the action of hormones in a living body. When taken into a living body, hormone action, such as the synthesis, storage, and secretion of hormones, is obstructed. For example, EDCs can inhibit generative functions, possibly causing malignant tumors.

7.5.1   Nonylphenol

The statistically significant difference was taken as the NOEC of 0.1 mg/L of Nonylphenol, and there was a statistically significant difference observed between the experimental microcosm and the control system with the addition of 1 mg/L. Therefore, it was determined there was influence on the ecosystem with the addition of 1 mg/L of Nonylphenol. The amounts of production (P) and respiration (R) at a Nonylphenol concentration of 0.2 mg/L were nearly the same as those of the control system, but there was a slight difference observed between concentrations of 0.1 mg/L and 0.2 mg/L (Fig. 7.21).

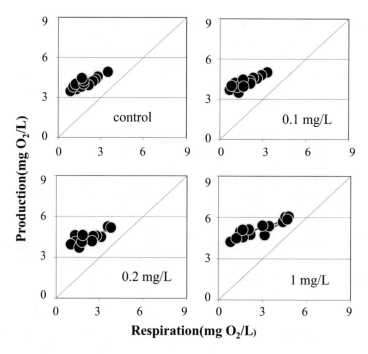

Fig. 7.21 Time course of P/R ratio in Nonylphenol-added microcosm N-system

7.5.2 Dibutyl Phthalate

Dibutyl phthalate is an organic compound, which is widely used as an additive in adhesives, print inks, and plasticizers. With the oily liquid, which has a sweet smell, alcohol, ether, and benzene are used as organic solvents. Whether this compound causes disturbance to the internal secretion of hormones is doubted. The effect of Dibutyl phthalate was investigated by Dr. Sugiura and added to the microcosm at the start of cultivation (day 0), and the populations of microorganisms and the amounts of respiration and production were tracked continuously. The m-NOEC of Dibutyl phthalate was estimated to be 1 mg/L (Sugiura 1992).

7.6 Antibiotics

Antibiotics, also called antibacterials, are a type of antimicrobial drug used in the treatment and prevention of bacterial infections. They may either kill or inhibit the growth of bacteria. A limited number of antibiotics also possess antiprotozoal capabilities. Antibiotics are not effective against viruses, such as the common cold or influenza; drugs that inhibit viruses are termed "antiviral drugs" or "antivirals," rather than "antibiotics." Antibiotics revolutionized medicine in the twentieth century. However, their effectiveness and easy access have also led to their overuse, prompting bacteria to develop resistance. This has led to widespread problems and prompted the World Health Organization (WHO) to classify antibiotic resistance as a "serious threat is no longer a prediction for the future, it is happening right now in every region of the world and has the potential to affect anyone, of any age, in any country."

The appearance of drug-resistant bacteria, such as methicillin-resistant *Staphylococcus aureus* (MRSA), is a problem resulting from the diversity and abuse of antibiotics, and multiple-drug-resistant bacteria, called "super resistant bacteria," began appearing in 2010, quickly becoming a major societal problem. Antibiotics are used not only in medical care but also in stock raising (i.e., ranching) and the aquaculture business, particularly when mixed with bait and supplied in nurseries. Not only are antibiotics discharged as excrement, but they are also excreted from bait directly into the environment and may have a great impact on aquatic ecosystems. Therefore, it is important to understand the influences of antibiotics, including the appearance of drug-resistant bacteria, on aquatic ecosystems.

7.6.1 Oxytetracycline

The antibacterial spectrum is wide, and the antibiotic Oxytetracycline (OTC) is used in the marine products industry, most commonly including aquaculture, and resistant bacteria have been confirmed in the outdoors. However, there has been no ecosystem impact assessment of OTC reported. The test results of oxytetracycline were compared with 43 ecological influence tests (42 single-species examinations and a large-scale outdoor mesocosm). The LOEC of the large-scale outdoor microcosm was estimated to be 100 µg/L, whereas the m-NOEC of the microcosm test was 70 µg/L, and it was demonstrated that the microcosm test is sufficiently applicable for use as an OECD test. Additionally, it succeeded in determining the relative community metabolism (RCM) (i.e., the metabolic activity of the community), and a difference was observed in the metabolic activity as a result of exposure to OTC. It was also shown that the microcosm test was able to reflect the mechanisms of the antibiotic influence. Furthermore, because RCM changed with changes in the community structure, what a near menstruation state could reproduce was suggested by impossible natural ecosystem in the conventional test method, which did not include community structure, and was able to show the superiority of the microcosm test for this assessment.

Resistant bacteria appeared in the microcosm by adding the antibiotic OTC. The shape of the bacterial colony cultured in the control system in a PY nutrient medium was different from the form of the bacterial colony cultured in the nutrient medium and which had OTC, clearly to become 25 mg/L to PY nutrient medium. The colony pro-control was whitish as a whole, and the border of the colony was not clear. Two colonies, which greatly varied in shape, appeared in the system with the addition of 7 mg/L of OTC. The first colony was generally light yellow and viscous, and the second colony was white, and the border exhibited clear granulation. Because the latter colony did not appear in the control system cultured in the PY nutrient medium, these were judged to be OTC-resistant bacteria. The test lasted for 14 days after OTC was added to the system at concentrations ranging between 0.007 mg/L and 7 mg/L. No difference was recognized in the algal population, which was evaluated from the total number of bacteria and the density of chlorophyll *a* in comparison with the control system. Additionally, the population of *Philodina erythrophthalma*, a metazoan, was not affected, the same as with algae. However, an influence was observed in the populations of OTC-resistant bacteria, *Cyclidium glaucoma*, *Lecane* sp., and *Aeolosoma hemprichi*, according to the OTC concentration over time.

The influences observed for 14 days after the addition of OTC were categorized as acute (2–4 days later), subacute (7–10 days later), and chronic (14 days later) to investigate changes in the microbial population in detail. It was determined that the system was affected when a population increased or decreased beyond the range of the control population. Because the population of *Aeolosoma hemprichi* exhibited an acute decrease with the addition of 7 mg/L of OTC, the NOEC of *Aeolosoma hemprichi* was estimated as less than 7 mg/L. Additionally, the NOEC for each was judged to be under 0.007 mg/L because increases and decreases in the populations of *Cyclidium glaucoma* and *Lecane* sp. were observed following the addition of 0.007 mg/L of OTC. Additionally, the population of *Cyclidium glaucoma* increased with OTC addition, while the population of *Lecane* sp. decreased. Because a common organism (bacteria) is assumed as the prey in this test, it is thought that the observed population changes are an outcome of the competitive interactions between organisms in the microcosm. Moreover, the NOEC of OTC-resistant bacteria was estimated to be less than 0.07 mg/L because an increase was observed in a number of bacteria with the addition of 0.07 mg/L of OTC. In this way, it is possible to estimate the NOEC for subacute and chronic influences equally. The effects of OTC on algae, all bacteria, and *Philodina erythrophthalma* were minimal, and the NOEC of each organism was greater than 7 mg/L for acute, subacute, and chronic influences. On the other hand, *Cyclidium glaucoma* and *Lecane* sp. were susceptible to OTC, and the NOEC for these organisms was less than 0.007 mg/L, except for the subacute influence (0.07 mg/L or less) of *Lecane* sp. It became clear from the population changes observed in the microcosm that the NOEC was less than 0.007 mg/L.

When high concentrations of OTC were added (7 mg/L and 0.7 mg/L), the amounts of production (P) and respiration (R) decreased in comparison with the control system (0 mg/L) over time, and the decrease in the amount of respiration was particularly large (Fig. 7.22). It is unclear why the addition of OTC decreased the amount of respiration, but it is likely that OTC binds to the intracellular ribosome, decreasing metabolic activity. When low concentrations of OTC were added (0.07 mg/L and 0.007 mg/L), a temporary increase in the amount of respiration was observed in the system with 0.07 mg/L of OTC, but the amounts of production and respiration both showed a tendency to decrease, the same as in the high-concentration systems. However, there were fewer decrements than in the high-concentration systems. The decrease in the amounts of production and respiration

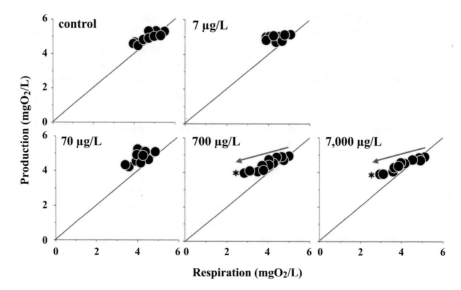

Fig. 7.22 Time course of P/R ratio in OTC-added microcosm N-system

was greater than in the control system in the lowest concentration (0.007 mg/L) system, and the m-NOEC estimated by changes in the P/R ratio was 0.007 mg/L.

The estimated NOEC was compared with the conventional value from the change in the populations mentioned above and the change in the P/R ratio. Based on the evaluation of population succession, the NOEC (greater than 7 mg/L) for the algae in this microcosm was larger than the conventional one column legal value (less than 0.11 mg/L). In the microcosm, the NOEC values for *Cyclidium glaucoma* and *Lecane* sp. were particularly low, at less than 0.007 mg/L. Evaluation by the US Environmental Protection Agency (EPA) has not been performed for these organisms, but the value (less than 0.007 mg/L) provided for the microcosm is less than the one column NOEC level judged in single-species tests of the EPA (less than 0.11 mg/L). Furthermore, the NOEC obtained from the P/R ratio was less than 0.007 mg/L, and it was revealed that it was susceptible to the addition of OTC into the microcosm.

The NOEC level determined above through the microcosm test (less than 0.007 mg/L) is less than the value (less than 0.11 mg/L) set by the US-EPA, and it is thought that this is because of the direct influence on the sensitivity of organisms to OTC, which is not considered in the EPA examination, and through the indirect influence of the interactions among organisms that were evaluated. The NOEC of OTC was estimated to be less than 0.007 mg/L, and it was suggested that the aquatic ecosystem may be influenced. In particular, it is possible that the addition of OTC, which influenced susceptible protozoans and metazoans, spread through the ecosystem through interactions among the constituent organisms.

7.7   Algal Toxins

In the aquatic ecosystem, pollution with microcystin, which are the toxins produced by high fecundity, eutrophying blue-green algae, combined with anthropogenic eutrophication, presents a major environmental problem. In eutrophic lakes during the summer season, blue-green algae such as *Microcyctis* multiply irregularly to large quantities, modifying the quality of the water and the biota of the aquatic ecosystem every year. In particular, blue-green algae produce toxins such as microcystin, and only very small amounts of these toxins that are present in lakes are derived from nature. However, an impact assessment of the aquatic ecosystem, especially the microbial ecosystem, with respect to contamination by microcystin, has not been performed. Therefore, evaluation of the influence of the toxin microcystin on microbial ecosystems, through examination of the biological inter-actions and material circulation among organisms and through the use of reduction functions, is important.

7.7.1   Microcystin-LR

Based on the structural parameter (population density), there was no difference observed between the populations of two microcystin-LR (where L stands for leucine and R for arginine) systems (0.1 mg/L and 1 mg/L addition) and the control system over time. The populations of *Chlorella* sp. and *Scenedesmus quadricauda*, both chlorophytes, remained stable between 10^5 N/mL and 10^6 N/mL after the addition of microcystin-LR. The population of *Tolypothrix* sp., a cyanophycean, was also stably maintained between 10^4 N/mL and 10^6 N/mL. As for the microanimals, the population of *Cyclidium glaucoma* decreased after the addition of microcystin-LR, but because this behavior was also observed in the control system, it was thought to simply reflect a natural change in the microcosm, and the population was stable between 10 Nm/L and 10^2 N/mL. Additionally, the populations of *Philodina erythrophthalma* and *Aeolosoma hemprichi* were stable and ranged from 1 N/mL to 10 N/mL.

With the addition of 0.1 mg/L and 1 mg/L of microcystin, it became clear that there was no remarkable change in the populations of this microcosm. Similarly, no remarkable changes in the P/R ratio between the two addition systems (0.1 mg/L and 1 mg/L of microcystin-LR addition) and the control system, which involved the addition of 480 mg/L of methanol, were recognized. However, there was a tendency for the amount of respiration (R) to increase in comparison with the additive-free control system, but it became clear that microcystin-LR did not have a remarkable influence on the P/R ratio under the conditions of this experiment (Fig. 7.23).

From these results, the m-NOEC of microcystin-LR was estimated to be more than 1 mg/L. This value is larger than 1 ppb, which is the guideline for concentrations in tap water established by the World Health Organization. It was made clear that microcystin-LR does not have a significant influence on the microcosm.

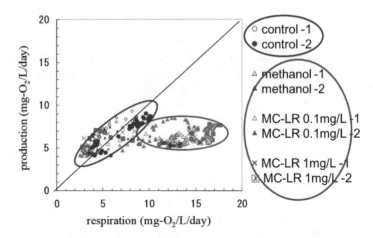

Fig. 7.23 P/R ratio in microcystin-LR-added microcosm N-system
O: without methanol as solvent for microcystin-LR
O: with methanol (480 mg/L) as solvent for microcystin-LR

7.8 Organic Matter and Organic Solvents

The effect of organic loading of the flask-sized microcosm system was investigated and assessed from the succession pattern of microbiota, the abundance in the stable state (N_{30}), and the biomass in the organic-loaded period (B_{16-30}) to obtain basic information from bioremediation, which uses certain essential microorganisms to improve polluted environments.

7.8.1 Polypeptone

Polypeptone, as a bacterial nutrient source, was added to the microcosm at a concentration of 5 g/L. The bacterial population density increased with absorption of polypeptone, and the culture medium was colored yellow by the 2nd day. The density of CFUs of bacteria increased to 10^9 N/mL, and CO_2 emissions also increased. *Chlorella* sp. and *Tolypothrix* sp. decreased in their population sizes immediately. Microanimals, such as *Cyclidium glaucoma* and *Philodina erythrophthalma*, became extinct, and only bacteria survived in the flask. The rate of succession in the microcosm increased in the 0.1 g/L addition system in comparison with the 0.5 g/L addition system.

Polypeptone, the substrate of the bacteria in the microcosm, was added as an organic matter load 16 days after the start of microcosm culturing. The addition quantity (quantity of the load) was assumed to be one, two, five, or ten times (i.e., $1\times, 2\times, 5\times$, or $10\times$) the quantity of existence in the microcosm on the 16th day after culturing began. In the $1\times$ polypeptone loading system, bacteria multiplied rapidly after the addition of polypeptone, and the protozoan, *Cyclidium glaucoma*, multiplied afterward. Then, the chlorophyceans, *Chlorella* sp. and *Scenedesmus quadricauda*, and the rotifers, *Lecane* sp. and *Philodina erythrophthalma*, multiplied in a more moderate fashion. In the 10x addition system, bacteria multiplied promptly after addition, and *Cyclidium glaucoma* multiplied but then suddenly decreased and perished afterward. Additionally, *Chlorella* sp., *Scenedesmus quadricauda*, *Lecane* sp., and *Philodina erythrophthalma* multiplied.

From the behaviors of the constituent organisms in the microcosm, it was observed that bacteria multiplied when organic matter was added and *Cyclidium glaucoma* multiplied as it preyed upon the bait. However, when organic matter was overloaded, because *Cyclidium glaucoma* disappeared after an increase in the concentration of organic matter, it was thought that the primary consumers (protozoans) were damaged (i.e., disappeared from the system) by too high of an addition concentration of the substrate. *Chlorella* sp. multiplied well, as evaluated by the comparison with the population (N_{30}) of the stable stage, but *Cyclidium glaucoma* perished at concentrations of greater than 250 mg/L. Organisms were shown to multiply in proportion to the increase in the concentration of polypeptone, except for the protozoan, *Cyclidium glaucoma*, and it became clear that the sensitivity of primary consumers (protozoans) to organic matter was extremely high. The rotifer, *Lecane* sp., followed by the protozoan, *Cyclidium glaucoma*, reacted sensitively based on a comparison of the population sizes before and after the addition (B_{16-30}) of polypeptone, and the possibility that *Lecane* sp., a bacterivore, participated as a buffer was demonstrated.

7.8.2 Acetone

When acetone is used as a solvent for a pesticide that is added to the microcosm, more than three times the respiration and production amounts are measured in the acetone addition system than are measured in the control system. Treated statistically (via ANOVA), there was an observed difference in the amounts of respiration and production between the control system and the solvent addition system on each measurement day. In the case of acetone addition, significantly higher values relative to the control system were shown for the respiration amount on the 9th day after addition and the production amount on the 7–9th days after addition. In this way, because the pesticide had very low solubility in water, it was necessary to dissolve it in organic solvents, such as acetone, but it became clear that the organic solvent itself caused an increase in the amounts of production and respiration.

7.8.3 Ethanol

When ethanol is used as a solvent for a pesticide that is added to the microcosm, more than three times the respiration and production amounts are measured in the acetone addition system than are measured in the control system. Treated statistically (via ANOVA), there was an observed difference in the amounts of respiration and production between the control system and the solvent addition system on each measurement day. In the case of acetone addition, significantly higher values relative to the control system were shown for the respiration amount on the 9th day after addition and the production amount on the 7–9th days after addition. In this way, because the pesticide had very low solubility in water, it was necessary to dissolve it in organic solvents, such as acetone, but it became clear that the organic solvent itself caused an increase in the amounts of production and respiration.

7.8.4　Methanol

When methanol is used as a solvent for a pesticide that is added to the microcosm, more than three times the respiration and production amounts are measured in the acetone addition system than are measured in the control system. Treated statistically (via ANOVA), there was an observed difference in the amounts of respiration and production between the control system and the solvent addition system on each measurement day. In the case of acetone addition, significantly higher values relative to the control system were shown for the respiration amount on the 9th day after addition and the production amount on the 7–9th days after addition. In this way, because the pesticide had very low solubility in water, it was necessary to dissolve it in organic solvents, such as acetone, but it became clear that the organic solvent itself caused an increase in the amounts of production and respiration.

7.8.5　Phenol

The effect of phenol was investigated and reported by Dr. Sugiura. Phenol was added at the time microcosm cultivation began (day 0), and the microorganism population and amounts of respiration and production were tracked over time. The amount of production became larger than that of the control system, but, with the addition of phenol, the amount of respiration was not observed to change with the addition of phenol when compared with the control system. The m-NOEC of phenol was estimated to be 5 mg/L (Sugiura 1992).

7.8.6　2.4-Dichlorophenol

2,4-Dichlorophenol is used as an organic, phosphorus-based insecticide and steril-izer, a phenoxy-based weed killer, and for herbicidal raw materials, and it may be generated by the resolution of a pesticide. The effect of 2,4-dichlorophenol was investigated and reported by Dr. Sugiura, and it was added to the microcosm when cultivation began (day 0); the microorganism population and amounts of respiration and production were tracked over time. The m-NOEC of 2,4-dichlorophenol was estimated to be 10 mg/L (Sugiura 1992).

7.9 Nutrients

7.9.1 Phosphate

An ecological impact assessment of PO_4-P on the P/R ratio in an experimental microcosm system was performed. The endpoints were abundance (structural parameter) and the concentration of DO (functional parameter), and the population was measured using an optical microscope and counted from the start of culturing on days 0, 2, 4, 7, 14, 16, 18, 20, 23, and 30, and it was evaluated from the results of B_{16-30} (days 16–30), which was the ratio of abundance and population density (N_{30}) on the 30th day. The DO concentration was measured continuously from the 16th day, and the P/R ratio was calculated from the amounts of production (P) and respiration (R).

On the 16th day after microcosm cultivation began, 30 mg/L of phosphate, which is a nutrient source for *Chlorella* sp. and *Tolypothrix* sp., was added. This is a very high concentration in comparison with that of natural aquatic ecosystems. Phytoplankton in the microcosm performed photosynthesis and absorbed phosphate, and the culture medium was colored a deep green. *Chlorella* sp. increased to 50 times that of its original population size, and CO_2 absorption and O_2 emission increased accordingly. Bacterial abundance increased simultaneously, whereas zooplankton, such as *Cyclidium glaucoma* and *Philodina erythrophthalma*, became extinct, and their corpses were observed.

7.9.2 Ammonia

The effect of ammonia was investigated and reported by Dr. Sugiura. Ammonia was added when microcosm cultivation began (day 0), and the populations of microorganisms and amounts of respiration and production were tracked over time. The amount of production became greater than that of the control system, but, with the addition of ammonia, the amount of respiration was not observed to change in comparison to the control system. The m-NOEC of ammonia calculated from the test concentration was 0.35 mg/L (Sugiura 1992).

7.10 Metals

Metal is worldwidely used for convenience in our life. On the other hand, it caused serious problem in the society such as Minamata disease by exposure of organic mercury (Hg) and Itai-itai disease by exposure of cadmium (Cd). In addition, utilization of silver nanoparticles (AgNP) as antibacterial agents has become a new environmental problem recently because of its environmental standard which has not been regulated yet.

Because heavy metals are typically highly toxic and are not biodegradable, they can easily remain in aquatic environments.

7.10.1 Cu

Assessment data for Cu by a microcosm test were reported by Dr. Sugiura. It was clear that the system did not collapse with the addition of 1.2 mg/L of Cu during the stable stage, although it collapsed with the addition of 0.4 mg/L during the initial stage of succession. The bioactivity in the microcosm was high in the initial stage and low in the stable stage of succession, indicating that the excretion amount is also high in the initial stage and low in the stable stage. The toxicity of Cu was expressed by bonding with organic matter in the microcosm. According to this, microorganisms survived, and the system was maintained in the stable stage, even with high concentrations of Cu being added. Organisms are susceptible to disturbance under high activity conditions, while they are strong under low activities. It was suggested that the stability of the system was maintained under coexisting and mutually poor conditions. The m-NOEC of Cu was estimated to be 0.16 mg/L (Sugiura 2001, 2009).

7.10.2 Zn

The effect of zinc was investigated by Dr. Sugiura, and it was added when microcosm cultivation began (day 0); the populations of microorganisms and amounts of respiration and production were tracked over time. The amount of production became larger than that of the control system, but, with the addition of Zn, the amount of respiration was not observed to change relative to the control system. The P/R ratio was 1 when 19.2 mg/L of Zn was added and became completely over-respirated in the 153.3 mg/L addition system, and the system collapsed. The m-NOEC of Zn was estimated to be 2.4 mg/L (Sugiura 2009) (Fig. 7.24).

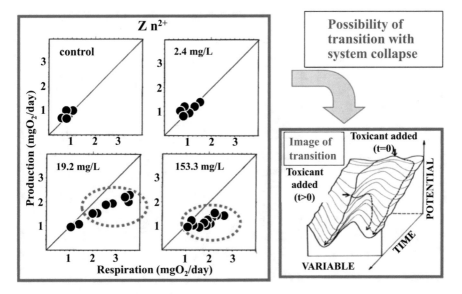

Fig. 7.24   Time course of P/R ratio in Zn-added microcosm N-system

7.10.3   Al

The concentrations of Al were 0.3, 0.6, 1.25, and 5.0 mg/L at the start of microcosm cultivation (day 0), and Al^{3+} was added by adjusting the concentration of $AlCl_3$. The initial concentration of Al^{3+} addition except 0.5–1.0 mg/L of systems; with respect to the population's metabolic rate, a change was observed over time that differed from the behavior of the control system. In particular, the concentration of Al^{3+} passed at 5 mg/L, and the state of the time change was different, and the extinction of organisms in the system was observed. The P/R ratio was 1 when 0.6 mg/L of Al was added, and it became completely over-respirated with the addition of 1.25 mg/L of Al, and the system collapsed (Fig. 7.25). The m-NOEC of Al was estimated to be 0.3 mg/L. These macrocosm tests were conducted by Dr. Sugiura (Sugiura 2001, 2009).

7.10.4   Cd

The effect of cadmium was investigated and reported by Dr. Sugiura. Cadmium was added at the time when microcosm cultivation began (day 0), and the populations of microorganisms and amounts of respiration and production were tracked over time. With the addition of Cd, the amounts of respiration and production became larger than in the control system. The m-NOEC of Cd was estimated to be 0.5 mg/L (Sugiura 2009).

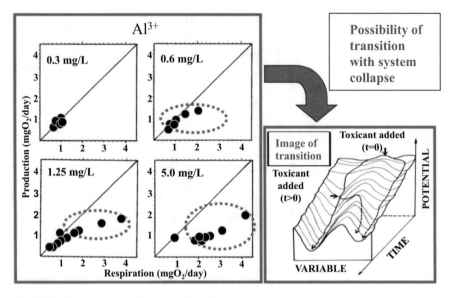

Fig. 7.25 Time course of P/R ratio in Al-added microcosm N-system

7.10.5 Mn

The addition concentration of Mn was adjusted to 0.1, 0.25, 0.5, 1.0, and 5.0 mg/L, and it was added on the 16th day of the stable state into the microcosm. The endpoints were abundance (structural parameter) and the concentration of DO (functional parameter), and the population was measured using an optical microscope and counted from the start of culturing on days 0, 2, 4, 7, 14, 16, 18, 20, 23, and 30; it was evaluated from the results of B_{16-30} (days 16–30), which was the ratio of abundance and population density (N_{30}) on the 30th day. The DO concentration was measured continuously from the 16th day, and the P/R ratio was calculated from the amounts of production (P) and respiration (R).

An evaluation of the risk of Mn on the structural parameter was conducted by comparison with the behavior of microorganisms in the control system 14 days after the addition of Mn. It was clear that the protozoan, *Cyclidium glaucoma*, was greatly affected at a concentration of 1.0 mg/L of Mn from the behaviors of representative species in the microcosm (Fig. 7.26). In other words, it was shown in the multi-species system that the influence varied according to microbial species. In N_{30}, the microbial community was divided into two groups, species that increased in abundance (microalgae) and those that decreased in abundance (microanimals). These groups were divided into two as N_{30} in B_{16-30}. Using the functional parameter, evaluation of the Mn risk was conducted in comparison with the behavior of the P/R ratio in the control system 14 days after the addition of Mn. An influence was observed at a concentration of 1.0 mg/L of Mn, especially in the respiration amount,

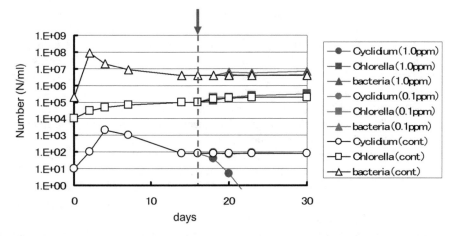

Fig. 7.26 Time course of population (representative species) in Mn-added microcosm N-system

and the P/R ratio decreased but recovered to ~1 (i.e., in the 1.0 mg/L Mn addition system, temporary toxicity occurred, but it was not chronic).

The m-NOEC of Mn was less than 1.0 mg/L, and a high sensitivity of the microcosm test was shown in comparison with the single-species test. The effectiveness of the assessment using the P/R ratio of the microcosm was indicated because of its similarity to the patterns of organismal abundance. However, further investigation of the m-NOEC calculation is needed.

7.10.6   Mg

The risk assessment of Mg as a sediment remediation material in aquatic ecosystems was conducted using a flask-sized microcosm system, based on the P/R ratio and the succession of microbial biota in comparison with the control system. The impact assessment of Mg was estimated from both the P/R ratio, as the functional parameter, and microbiota, as the structural parameter, in the microcosm. The environmental impact and ecological risk of these materials were estimated from comparison with the no-addition system (control) in both assessment methods described below. The P/R ratio was calculated from the amount of oxygen produced during the daytime, the production rate, and the amount of oxygen consumed at night, the respiration rate. When the value of the P/R ratio is near 1, the ecosystem is in a stable state, but when the value is far from 1, the ecosystem is in a state of succession or collapse (Odum 1983; Odum and Horne et al. 1983). In the case in which the P/R ratio is ~1, the impact risk of Mg was assessed as not serious, but a risk is involved for cases in which values are far from 1. The state of the microbiota was estimated from (1) the succession pattern of each microorganism, (2) the abundance at 30 days after cultivation began (N_{30}), and (3) the biomass during the period from the addition of

Mg to the end of cultivation (B_{16-30}). The regular succession pattern is similar. After cultivation began, bacteria grew rapidly using a polypeptone in the basal medium, and after that the protozoan *Cyclidium glaucoma* predated these bacteria and increased its abundance. The chlorophyceans, *Chlorella* sp. and *Scenedesmus quadricauda*, grew via photosynthesis under artificial light. After that, the other microorganisms increased their numbers through microbial interactions, such as prey-predator interactions, and the succession of microbiota into the stationary phase 14 days after cultivation began. The succession pattern of the P/R ratio in the period from the day of Mg addition to the final day of microcosm cultivation was compared with the control system. The results showed that the respiration rate (R rate) was greatly affected in the case of a 10.0 mg/L concentration of Mg injection in comparison with the production rate (P rate) (i.e., the respiration rate increased from 0.9 mg/day on the 16th day to 4.3 mg/day on the 18th day (maximum) and recovered to ~1.1 mg/day by the 24th day). Similarly, the production rate increased from 1.2 mg/day on the 16th day to 2.2 mg/day on the 18th day (maximum) and recovered to ~1.1 mg/day by the 24th day. The P/R ratio temporarily decreased from 1 to 0.5, and, after that, the P/R ratio recovered to ~1. Based on the results from statistical analyses, there was a significant difference between the system with 10.0 mg/L of Mg added and the control ($p < 0.05$) regarding the R rate in the period between the 16th day and the 24th day and the P/R ratio during the period between the 16th day and the 21st day. Conversely, there was no significant influence on the P/R ratio observed at a Mg concentration of 5.0 mg/L ($p < 0.05$). From this, the 10.0 mg/L concentration of Mg was considered toxic, although not chronic (but temporary), with respect to the P/R ratio as the functional parameter.

From an evaluation of the succession pattern of microorganisms as the structural parameter, the protozoan, *Cyclidium glaucoma*, and oligochaete, *Aeolosoma hemprichi*, were greatly influenced by the addition of 10.0 mg/L of Mg, and these two species decreased in their abundance in the 10.0 mg/L addition system. Other microanimals, such as the rotifers, *Lecane* sp. and *Philodina erythrophthalma*, also decreased in comparison with the control system. Thus, the influence of Mg differed for different species under coexisting culture conditions, such as those in the microcosm. Conversely, there was no significant difference observed in the abundance of organisms, when compared with the control system, at Mg concentrations of under 5.0 mg/L. As for the water quality, the pH rose with the addition of Mg, but the value of the pH was approximately 8.5–9.5 in all experiments. From an estimation of N_{30} at a concentration of 10.0 mg/L of Mg, the microorganisms were divided into two groups, those with increasing abundance and those with decreasing abundance. The former group contained algae as the producers (*Chlorella* sp., *Scenedesmus quadricauda*, and *Tolypothrix* sp.) and bacteria as the decomposers (*Pseudomonas putida*, *Acinetobacter* sp., *Bacillus cereus*, coryneform bacteria, etc.), and the latter group contained protozoans and metazoans as the consumers (*Cyclidium glaucoma*, *Lecane* sp., *Philodina erythrophthalma*, and *Aeolosoma hemprichi*). From an estimation of B_{16-30}, the microorganisms were divided into two groups, the increasing abundance and the decreasing abundance groups, the same as for N_{30}. The m-NOEC of Mg was estimated to be more than 10.0 mg/L.

7.10.7   Ca

An ecosystem impact assessment of Ca was conducted using both the P/R ratio as the functional parameter and the microbiota as the structural parameter in the microcosm. The environmental impacts and ecological risks of these materials were estimated from comparison with the no-addition system (control) in both assessment methods. The succession pattern of the P/R ratio during the period that began from the day Ca was added and lasted until the final day of microcosm cultivation was compared with that of the control system. The results showed that the P rate was more strongly affected at a 5.0 mg/L concentration of Ca in comparison with the R rate (i.e., the P rate decreased from 1.1 mg/day on the 16th day to 0.6 mg/day on the 21st day (maximum) and recovered to ~1.0 mg/day by the 24th day). Meanwhile, the R rate increased from 0.9 mg/day on the 16th day to 2.1 mg/day on the 18th day (maximum) and recovered to ~1.2 mg/day by the 23rd day. The P/R ratio temporarily decreased from 1.2 to 0.4, and, after that, it recovered to ~1 by the 23rd day. The microcosm was considered to be influenced by Ca concentrations of 5.0 mg/L and 10.0 mg/L in this study. According to the results of statistical analysis, there was a significant difference among the systems with 5.0 mg/L and 10.0 mg/L of added Ca and the control ($p < 0.05$). With the P/R ratio as the functional parameter, the 5.0 mg/L concentration of Ca was considered toxic, although not chronically. From an estimation of the succession pattern of microorganisms as the structural parameter, *Cyclidium glaucoma* and *Lecane* sp. were greatly influenced by the addition of 5.0 mg/L of Ca, and these two species decreased in their abundance with the addition of 5.0 mg/L of Ca, but other microanimals, such as the rotifer, *Philodina erythrophthalma*, decreased in comparison with the control microcosm. From an estimation of N_{30} at a concentration of 5.0 mg/L of Ca, the microorganisms were divided into two groups, those of increasing abundance and decreasing abundance. The former group contained bacteria as the decomposers, and the latter contained protozoans and metazoans as the consumers. Algae, as the producers, were divided into two groups; the abundance of the chlorophyceans, *Chlorella* sp. and *Scenedesmus quadricauda*, decreased, and the abundance of cyanophycean, *Tolypothrix* sp., increased. This phenomenon was caused by the growth inhibition of microanimals, especially *Aeolosoma hemprichi*, by the sudden increase in pH (from 8.2 to 11.1 just after the addition of 5.0 mg/L of Ca) and the photosynthetic inhibition of microalgae, especially *Chlorella* sp. and *Scenedesmus quadricauda*, by the light inhibition of Ca. The oligochaete, *Aeolosoma hemprichi*, has a strong jaw for eating and destroying flocs contained in bacteria, algae, other organisms, and detritus (Inamori et al. 1990). It was thought that *Aeolosoma hemprichi* was damaged in its physiological activity by the increase in pH; subsequently, *Tolypothrix* sp. escaped from the predation by *Aeolosoma hemprichi*. Under microscopic observation, it was confirmed that *Aeolosoma hemprichi* predated flocs in *Tolypothrix* sp., bacteria, and detritus in the control microcosm. The coefficients of variation (standard deviation/mean, expressed as percentages) for the abundance of each microorganism in the microcosm during the cultivation period were 10% or less (50 independent experiments). This value is almost the same as previously reported values (e.g., Inamori et al. 1992; Kurihara 1978a, b).

From these outcomes, the effects of Mg and Ca on the aquatic ecosystem as sediment remediation materials both indicate powerful nutrient removal (especially for phosphorus) by bonding with PO_4-P through chemisorption in the water eluted from the eutrophied sediment (Murakami et al. 2000), differing in both influential concentrations and target microorganisms. The impact risk to the aquatic ecosystem of Mg was estimated as half that of Ca from the viewpoint of the concentration that had any influence on the microorganisms in the microcosm. The m-NOEC was estimated to be less than or equal to 10.0 mg/L for Mg and as less than or equal to 5.0 mg/L for Ca, from the viewpoint of changes in the P/R ratio, which is considered to be the functional parameter of the microcosm. From the biotic succession, as the structural parameter, the m-NOEC was considered to be less than or equal to 5.0 mg/L for Mg and much lower than 5.0 mg/L for Ca, with the abundance of protozoans, rotifers, and oligochaetes obviously decreasing and not recovering. However, the P/R ratio of the microcosm was maintained at the value of 1, notwithstanding the different biota, which consisted of producers, consumers, and decomposers, as the basic components of the ecosystem. Considering that the P/R ratio of the natural ecosystem in the stable state converges to almost 1 (Odum 1983), the m-NOEC should be estimated by the P/R ratio (functional parameter) rather than by the biotic succession (structural parameter). Further discussion and experimental data are necessary to better understand this estimation method.

7.10.8 Ni

Nickel is used for alloying and plating, and it may become mixed with tap water by elution from mine wastewater and nickel plating. The toxicity of Ni is different from a chemical form in the properties of matter; the oral toxicity is relatively low and is at the same level as the requisiteness of other metals, such as Cu, Co, and Zn, but the toxicity of inhaled nickel carbonyl is high, and the fatal dose for *Homo sapiens* has been estimated as 30 ppm/30 min. For Ni and its compounds, the targeted value for water quality management is less than 0.01 mg/L (temporarily).

Nickel was supplied for assessment, and the no-addition system (control) and addition systems (0.1, 1.0, 5.0, and 10 mg/L) were adjusted and added 16 days after culturing began. The endpoints were abundance (structural parameter) and the concentration of DO (functional parameter), and the population was measured using an optical microscope and counted from the start of culturing on days 0, 2, 4, 7, 14, 16, 18, 20, 23, and 30; it was evaluated from the results of B_{16-30} (days 16–30), which was the ratio of abundance and population density (N_{30}) on the 30th day. The DO concentration was measured continuously from the 16th day, and the P/R ratio was calculated from the amounts of production (P) and respiration (R).

As the result of the gnotobiotic-type microcosm test (N-system), after evaluating the ecosystem risk from the change in the P/R ratio as the functional parameter, no influence was observed at 0.1 mg/L of Ni addition, but respiration increased with 1.0 mg/L of addition, and the P/R ratio was greater than 1. From this, it was thought that Ni acted to increase the amount of respiration in the system. Microbiota in the microcosm were divided into two groups, those of increased abundance (microalgae) and those of decreased abundance (microanimals) in comparison with the control system, using the abundance as the structural parameter. The respiration activity *per* microanimal population rose, and the production activity *per* microalgae decreased with an increasing concentration of Ni, based on the evaluation of changes in the DO concentration *per* individual, because B_{16-30} revealed a decrease in the microanimal population with the addition of Ni.

From these outcomes, the m-NOEC of Ni was estimated to be between 0.1 and 1.0 mg/L. Additionally, as the result of the naturally derived microcosm test, the P/R ratio (functional parameter) converged to approximately 1, but the microbiota in the control (no-addition system) were not stable, and the abundance of organisms (structural parameter) also differed. Reproducibility was not maintained in all of the systems. During the culture period, *Chlorella* sp., *Nitzschia* sp., *Monoraphidium contortum*, and *Aulacoseira granulata* ordinarily appeared, but no characteristic behavior linked to Ni addition was observed. As the result of the stress-selected-type microcosm test, only 7 stable systems out of the 40 total systems were established. Even in the control system, the reproducible stability was insufficient. No remarkable change was observed, regardless of the concentration of the Ni addition, and the P/R ratio varied near 1 and was stable. However, a clear decrease in population occurred at a 1.0 mg/L concentration of Ni, and it was shown that *Nitzschia* sp. and *Monoraphidium contortum* were particularly affected. It was predicted that diatomaceans were affected more than chlorophyceans.

7.10.9 Co

Cobalt was supplied for impact assessment to the no-addition system (control) and addition systems (1, 1.5, 2, 4, 6, 8, and 10 mg/L concentrations) and was added 16 days after the start of culturing. The endpoints were abundance (structural parameter) and the concentration of DO (functional parameter), and the population was measured using an optical microscope and counted from the start of culturing on days 0, 2, 4, 7, 14, 16, 18, 20, 23, and 30; it was evaluated from the results of B_{16-30} (days 16–30), which was the ratio of abundance and population density (N_{30}) on the 30th day. The DO concentration was measured continuously from the 16th day, and the P/R ratio was calculated from the amounts of production (P) and respiration (R).

As a structural parameter, the acute influence B_{16-20} (0–4 days after loading), subacute influence B_{20-23} (5–7 days after loading), chronic influence B_{23-30} (8–14 days after loading), and B_{16-30} as abundance were estimated. As a result, microorganisms did not perish during the period of acute influence, but extinction

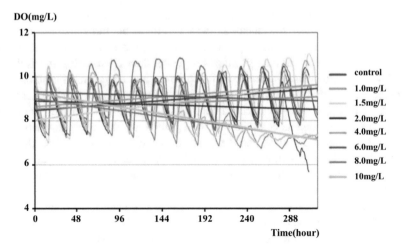

Fig. 7.27 Time course of DO in Co-added microcosm N-system

was observed during the periods of subacute influence and chronic influence. Additionally, it was revealed that many populations decreased in the high-concentration systems from B_{16-30}. As a functional parameter, the amplitude of the DO was damped by addition concentrations of more than 8 mg/L, and it is thought that the system collapsed (Fig. 7.27). The P/R ratios in all systems were stable at ~1. From these results, the m-NOEC of the structural parameter was estimated to be 1.5 mg/L of Co because microorganisms in the microcosm perished in more than 2 mg/L during the period of chronic influence. Meanwhile, the m-NOEC of the functional parameter was estimated to be 6 mg/L because the system collapsed with concentrations of more than 8 mg/L of Co from the viewpoint of changes in the DO concentration over time.

7.10.10   Ni and Co

Cobalt and Ni were supplied for assessment, and a no-addition system (control) and complexed addition systems (Co 0.2 mg/L + Ni 0.02 mg/L, Co 0.4 mg/L + Ni 0.04 mg/L, Co 0.6 mg/L + Ni 0.06 mg/L, Co 0.8 mg/L + Ni 0.08 mg/L, Co 1 mg/L + Ni 0.1 mg/L, Co 3 mg/L + Ni 0.3 mg/L, Co 5 mg/L + Ni 0.5 mg/L, and Co 10 mg/L + Ni 1 mg/L) were adjusted and loaded 16 days after culturing began. The addition concentration was set on the basis of the m-NOEC of Co and Ni. The endpoints were abundance (structural parameter) and the concentration of DO (functional parameter), and the population was measured using an optical micro-scope and counted from the start of culturing on days 0, 2, 4, 7, 14, 16, 18, 20, 23, and 30; it was evaluated from the results of B_{16-30} (days 16–30), which was the ratio of the abundance and population density (N_{30}) on the 30th day. The DO concentration was measured continuously from the 16th day, and the P/R ratio was calculated from the amounts of production (P) and respiration (R).

As the structural parameter, B_{16-30}, the abundance 14 days after the addition of Co, was evaluated. Microorganisms in the microcosm did not perish during the period of acute influence, but extinction was observed during the periods of subacute influence and chronic influence. Increases and decreases in the populations of each microorganism were observed with Co addition, and the tendency for populations to slightly increase in a low-concentration system was recognized in *Philodina erythrophthalma*, but large decreases in population were observed in *Cyclidium glaucoma* and *Lecane* sp. Additionally, it was revealed that many populations decreased in the high-concentration systems from B_{16-30}. It is thought that the population ratio changed due to differences in the Co-resistance of each microbial species. As a functional parameter, the amplitude of the DO damped in the 5 mg/L of Co + 0.5 mg/L of Ni system, and it is thought that the system collapsed. The P/R ratio in all systems was stable at approximately 1.

From these results, the m-NOEC of the structural parameter was estimated to be 1 mg/L of Co + 0.1 mg/L of Ni because microorganisms in the microcosm perished at this concentration during the period of chronic influence. Meanwhile, the m-NOEC of the functional parameter was estimated to be 5 mg/L of Co + 0.5 mg/ L of Ni because the system collapsed at concentrations higher than this from the viewpoint of the changes in the concentration of DO. In comparison with Co-only addition system, the m-NOEC was nearly equal, but the biomass ratio of the zooplankton largely decreased in the low-concentration system.

7.10.11   Cs

Cesium was supplied for assessment, and a no-addition system (control) and addition systems (0.1, 0.5, 1, and 2 mg/L) were adjusted and added as CsCl 16 days after the start of cultivation. The endpoints were abundance (structural parameter) and the concentration of DO (functional parameter), and the population was measured using an optical microscope and counted from the start of culturing on days 0, 2, 4, 7, 14, 16, 18, 20, 23, and 30; it was evaluated from the results of B_{16-30} (days 16–30), which was the ratio of the abundance and population density (N_{30}) on the 30th day. The DO concentration was measured continuously from the 16th day, and the P/R ratio was calculated from the amounts of production (P) and respiration (R).

As a structural parameter, although a decrease was observed in the population of the protozoan, *Cyclidium glaucoma*, which was the primary consumer, no major change was observed for the entire microcosm. However, *Cyclidium glaucoma* perished on the 30th day (14th day after addition) at a Cs concentration of 1 mg/L, and an immediate decrease in the population after addition was observed at a concentration of 2 mg/L. From this, the m-NOEC was estimated to be less than 1 mg/L for Cs. Additionally, it was thought that there was a possibility of metastasis of the system due to Cs addition because the microanimals, such as *Philodina erythrophthalma*, and microalgae, such as *Chlorella* sp., coexisted after *Cyclidium glaucoma* disappeared from the system and maintained a constant population. As a functional parameter, from the time series of the DO concentration in the microcosm, a 0.1 mg/L addition of Cs increased the activity and coped with the Cs load, and the 1 mg/L addition system recovered its activity through microbial interactions, material circulation, and energy flow among the microorganisms in the microcosm after the activity decreased due to the addition of Cs. Because the P/R ratio (functional parameter) is ~1, and the system was estimated to be stable, all addition systems were estimated as stable. From these results, with respect to the P/R ratio, the ecosystem function in the microcosm was judged to be stable when Cs addition concentrations were less than 1 mg/L. As a result of having analyzed the ecosystem influence of the Cs from the P/R ratio and microbial abundance (structural parameter), and because *Cyclidium glaucoma* was not observed, though the activity of the system was restored at a Cs concentration of 1 mg/L, it was thought that differential structuring of the system occurred.

From these outcomes, the m-NOECs of both the structural parameter and functional parameter were estimated to be less than 1 mg/L of Cs. Additionally, when it was assumed that a radiation of 3000 Bq was emitted from 1 mg of Cs based on the ^{137}Cs data from the Chernobyl nuclear power plant accident (Iimoto 2012; Minai et al. 2011), 0.02 mg of Cs existed in a 200 ml volume of microcosm culture fluid by the addition of 0.1 mg/L of Cs, and the half-life of ^{137}Cs was applied to calculate the type of radiation exposure dose in the microcosm (Iimoto 2012) for 30 years; the radiation exposure dose, D, was approximately 30 Sv. That is, in the microcosm, approximately 30 Gy/day would be irradiated at a Cs concentration of 0.1 mg/L. It was shown that there was a difference of ~10 times the radiation exposure (hot run) and m-NOEC of the metal load (cold run) in the microcosm when following the conversion mentioned above because there was some influence at the 23 Gy/day dose rate and almost no influence at the 10 Gy/day dos rate observed from the gamma beam irradiation experiment, and it was estimated that the radiation exposure had an influence that was approximately ten times as strong as that of the metal load. Here, the ecosystem influence of the addition of Cs was analyzed using a cold run, using the cold Cs, but it is difficult to convert values from Bq to Gy and Sv because they rely on different definitions for radioactive Cs, and further examination is necessary to complete the ecosystem impact assessment of Cs pollution.

7.10.12 I$_2$

Iodine was supplied for assessment, and a no-addition system (control) and addition systems (8 and 10 mg/L) were adjusted and added 16 days after the start of culturing. The endpoints were abundance (structural parameter) and the concentration of DO (functional parameter), and the population was measured using an optical microscope and counted from the start of culturing on days 0, 2, 4, 7, 14, 16, 18, 20, 23, and 30; it was evaluated from the results of B$_{16-30}$ (days 16–30), which was the

Fig. 7.28 Time course of
P/R ratio in 10 mg/L of
I_2-added microcosm
N-system

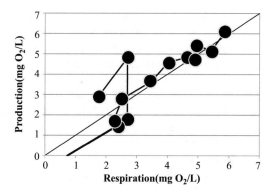

ratio of the abundance and population density (N_{30}) on the 30th day. The DO
concentration was measured continuously from the 16th day, and the P/R ratio
was calculated from the amounts of production (P) and respiration (R).

As a structural parameter, a decrease was observed at a concentration of 8 mg/L of
I_2 in the populations of *Cyclidium glaucoma*, *Lecane* sp., *Scenedesmus quadricauda*,
and *Tolypothrix* sp., but no microorganisms in the microcosm perished. Extinction
was observed for *Cyclidium glaucoma*, *Philodina erythrophthalma*, and *Aeolosoma
hemprichi* at an I_2 concentration of 10 mg/L, and the populations of *Scenedesmus
quadricauda* and *Tolypothrix* sp. also decreased. Additionally, the tendency for only
Lecane sp. to increase in comparison with other zooplanktons in N_{30} was recognized.
These were influences upon the microcosm due to the inhibition of photosynthesis by
I_2, and it was thought that extinction did not occur because the phytoplankton had a
high ability to acclimate. As a functional parameter, bioactivity was observed at an I_2
concentration of 8 mg/L, and the P/R ratio was relatively stable. Activity decreased
with the addition of 10 mg/L of I_2, and the P/R ratio measured just as this began was
unstable, but it stabilized after the 22nd day of culturing (Fig. 7.28). Both production
and consumption decreased, but the system remained stable. It was thought that the
microbial interactions were responsible for recovery of the system after it collapsed
under the influence of I_2. From these outcomes, it was estimated that the m-NOECs of
both the structural parameter and functional parameter were present in the range
between 8 mg/L and 10 mg/L because a tendency toward decreasing populations of
both producers and consumers was recognized, whereas the P/R ratio of the system
was recovered after the addition of I_2.

7.10.13 *Cs and I_2*

Cesium and I_2 were supplied for assessment, and a no-addition system (control) and
complexed addition systems (0.5 mg/L of Cs + 1.5 mg/L of I_2, 1.0 mg/L of
Cs + 3.0 mg/L of I_2, 1.5 mg/L of Cs + 4.5 mg/L of I_2, 2.0 mg/L of Cs + 6.0 mg/L
of I_2, 2.5 mg/L of Cs + 7.5 mg/L of I_2, 0.25 mg/L of Cs + 0.75 mg/L of I_2, 2.7 mg/L

of Cs + 8.0 mg/L of I_2, and 8.0 mg/L of Cs + 2.7 mg/L of I_2) were adjusted and loaded 16 days after culturing began. The addition concentrations were set on the basis of the m-NOECs of Cs and I_2. The endpoints were abundance (structural parameter) and the concentration of DO (functional parameter), and the population was measured using an optical microscope and counted from the start of culturing on days 0, 2, 4, 7, 14, 16, 18, 20, 23, and 30; it was evaluated from the results of B_{16-30} (days 16–30), which was the ratio of the abundance and population density (N_{30}) on the 30th day. The DO concentration was measured continuously from the 16th day, and the P/R ratio was calculated from the amounts of production (P) and respiration (R). As a result, the m-NOEC of the structural parameter was estimated to be 1.0 mg/L of Cs + 3.0 mg/L of I_2. In comparison with the I_2-only addition system, the *Aeolosoma hemprichi* population was recognized as having a tendency to increase.

7.10.14 Sr

Strontium was supplied for assessment, and a no-addition system (control) and addition systems (0.1, 0.3, 0.4, 0.5, 1, 4, 6, 7, 8, and 10 mg/L) were adjusted and added 16 days after culturing began. The endpoints were abundance (structural parameter) and the concentration of DO (functional parameter), and the population was measured using an optical microscope and counted from the start of culturing on days 0, 2, 4, 7, 14, 16, 18, 20, 23, and 30; it was evaluated from the results of B_{16-30} (days 16–30), which was the ratio of abundance and population density (N_{30}) on the 30th day. The DO concentration was measured continuously from the 16th day, and the P/R ratio was calculated from the amounts of production (P) and respiration (R).

As a structural parameter, after the addition of Sr, all of the microorganisms in the microcosm existed at a Sr concentration of 0.3 mg/L, but *Cyclidium glaucoma* perished with the addition of 0.4 mg/L. Additionally, other zooplanktons were observed to decrease in abundance with increased Sr concentrations. The population of the phytoplankton, *Scenedesmus quadricauda*, increased, and *Tolypothrix* sp. also increased from double to ~3.5 times its initial abundance. These species increased with higher concentrations of Sr. It was thought that Sr worked to decrease the abundance of zooplankton and to multiply phytoplankton. As a functional parameter (the P/R ratio), the activity increased and was not disturbed at concentrations of 7 mg/L, but it was unbalanced with the addition of 8 mg/L of Sr near the end of the 25th day. For the N30 of the 7 mg/L and 8 mg/L concentrations, phytoplankton increased in both concentrations. It was thought that there was a mechanism that caused reduction in zooplankton activity at high concentrations. Additionally, the P/R ratio collapsed on the consumption side at a concentration of 8 mg/L (Fig. 7.29). Because the abundance of zooplankton decreased, the multiplication of phytoplankton was affected and decreased in its activity, and it is thought that phytoplankton consumed more than they produced. From these outcomes, the m-NOEC of the structural parameter was estimated to be 0.3 mg/L, and the m-NOEC of the

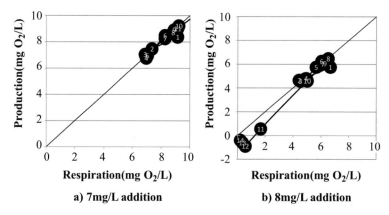

Fig. 7.29 Time course of P/R ratio in Sr-added microcosm N-system. (**a**) 7 mg/L addition. (**b**) 8 mg/L addition

functional parameter was estimated to be 7 mg/L. The toxicity of Sr affected zooplankton consumers, and the activity of phytoplankton decreased at high concentrations.

7.10.15 Silver Nanoparticles (AgNP)

In recent years, silver nanoparticles (AgNP) have come to be widely used as antibiotics and sterilizers, but information is not available on the influence these particles may have on aquatic ecosystems through drainage into river water after use. Though the metal (Ag) is known to exhibit serious toxicity in aquatic ecosystems that contain biological interactions, material circulation, and energy flow, the environmental risk remains poorly understood. The microcosm, which is the model microbial ecosystem consisting of producers, consumers, and decomposers, is useful for evaluating the environmental risk to an ecosystem system with complex interactions among organisms, rather than the toxic risk evaluation for single species at an ecosystem level, and positioning as a standard docimasy becomes important. Here, we conducted an environmental impact assessment of AgNP on microbial community function and structure, using the experimental flask-sized microcosm system.

To assess its influence, AgNP was added at several concentrations on the 16th day after microcosm cultivation began. Plankton observation using an optical microscope was conducted at the start of culturing, on days 0, 2, 4, 7, 14, 16, 18, 20, 23, and 30, and the kinds of changes in a flask for each addition concentration as the structural parameter were observed. In addition, DO was measured consecutively from the 16th day during the culture period to evaluate the function side by the P/R ratio as the function parameter.

The appearance of the microcosm succeeded from green in the non-addition system (control system) to a strong gray with the increasing addition concentration

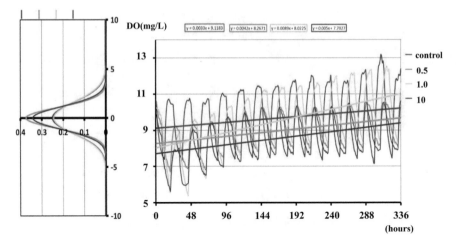

Fig. 7.30 Time course of DO in AgNP-added microcosm N-system

of AgNP, and a large influence was produced in the microcosm. To focus on the P/R ratio, the amplitude of the DO concentration tended to decline with increasing of AgNP concentrations, as shown in Fig. 7.30. The activity of the microcosm decreased with the addition of 0.5 mg/L of AgNP, and the activity of the system ceased with 10 mg/L of added AgNP. A large population decrease in the producers was not observed with the addition of AgNP for the chlorophyceans, *Chlorella* sp., and *Scenedesmus quadricauda*, but the population decreased and the individual lengths shortened for the cyanophycean, *Toplypothrix* sp. with AgNP addition. The populations of the consumers, *Cyclidium glaucoma* and *Aeolosoma hemprichi*, decreased at concentrations of 0.5 mg/L of AgNP, and the abundance of *Lecane* sp. slightly decreased in the 1.0 mg/L of AgNP addition system. Both *Cyclidium glaucoma* and *Aeolosoma hemprichi* perished, and the abundance of the rotifers, *Philodina erythrophthalma* and *Lecane* sp., decreased with the addition of 10 mg/L of AgNP (Fig. 7.31).

7.11   Irradiation

The lower-energy, longer wavelength part of the spectrum, including visible light, infrared light, microwaves, and radio waves, is nonionizing; its main effect when interacting with tissue is heating. This type of radiation only damages cells if the intensity is high enough to cause excessive heating. Ultraviolet radiation has some features of both ionizing and nonionizing radiation. While the part of the ultraviolet spectrum that penetrates the Earth's atmosphere is nonionizing, this radiation does far more damage to many molecules in biological systems than can be accounted for by the effects of heating, sunburn being a well-known example. These properties are

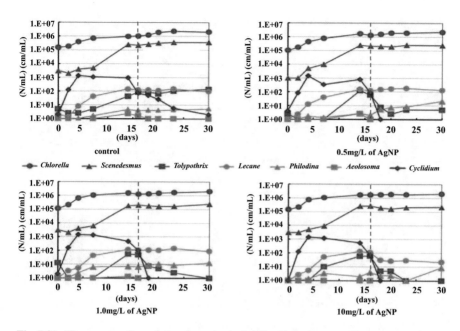

Fig. 7.31 Time course of population dynamics in AgNP-added microcosm N-system

derived from the ability of ultraviolet radiation to alter chemical bonds, even without having quite enough energy to ionize atoms.

Across the globe, the need for nuclear power generation is discussed as part of energy policies. The nuclear power plant accident during the earthquake disaster in eastern Japan, that is, the Tohoku earthquake and the resulting tsunami that affected power plants including Fukushima Daiichi, was the beginning of much discussion. With respect to nuclear power generation, the inspection and accumulation of knowledge about the safety for and effects of radiation on human beings, animals, plants, and the aquatic environment are necessary. The monitoring of pollution from fishery by-product pollution, soil pollution, and the accumulation of sewage grime was accomplished using radiological vegetables after the nuclear plant accident, and further analysis is now being pushed forward. However, consideration of the kind of influence radiation has on the small animals and phytoplankton constituting the base of the aquatic ecosystem, so-called microbes, remains limited. Here, the influence of radiation on the aquatic ecosystem is described.

7.11.1 γ-Ray (^{137}Cs)

To characterize the indirect effects of ionizing radiation on aquatic microbial communities, the effects of acute γ-irradiation were investigated in a microcosm (Fig. 7.32). Population changes in the constituent organisms were observed over

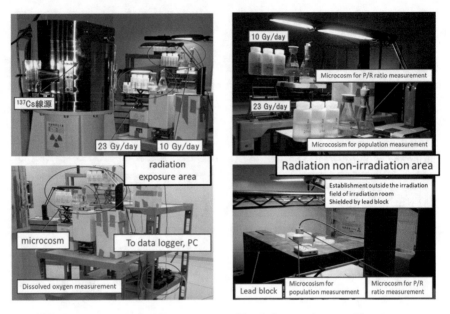

Fig. 7.32 Environmental impact risk assessment of irradiation on microcosm N-system

160 days after irradiation. The prokaryotic community structure was also examined by denaturing gradient gel electrophoresis (DGGE) of 16S rDNA. Principal response curve analysis revealed that the populations of the microcosm as a whole were not significantly affected at 100 Gy while they were adversely affected at 500–5000 Gy in a dose-dependent manner. However, some effects on each population, including each bacterial population detected by DGGE, did not depend on radiation doses, and some populations in the irradiated microcosm were larger than those of the control system. These unexpected results are regarded as indirect effects through interspecies interactions, and possible mechanisms are proposed originating from population changes in other organisms coexisting in the microcosm. For example, some indirect effects on consumers and decomposers likely arose from interspecies competition within each trophic level. It is also likely that predator-prey relationships between producers and consumers caused some indirect effects on producers.

The effects of acute γ-irradiation were investigated in the aquatic microcosm. At 100 Gy, populations were not affected in any taxa. At 500–5000 Gy, one or three taxa died out, and populations of two or three taxa decreased over time, while that of *Tolypothrix* sp. increased. This *Tolypothrix* sp. increase was likely an indirect effect due to interspecies interactions. Principal response curve analysis revealed that the main trend in the effects was a dose-dependent population decrease. For a better understanding of radiation risks in aquatic microbial communities, effect doses of γ-rays, compared with copper, herbicides, and detergents, were evaluated using a radiochemoecological conceptual model and the effect index for the microcosm. The populations of all microorganisms in the microcosm were stable under the irradiation conditions during an experimental period in a gamma beam irradiation experiment.

The cell count (CFU) of bacteria at 10 Gy/day (low dose of irradiation system) decreased more than in the control condition after irradiation began on the 19th day, but no influence was observed in other species. At 23 Gy/day (high dose of irradiation system), the abundance of *Tolypothrix* sp. increased to more than that of the control system after the 28th day. The bacteria decreased after the 4th day, and *Lecane* sp. decreased on the 45th day, but no influence was observed in other microorganisms. Culturing of the microcosm was conducted under a light and dark period for 12 h each, but, as for DO (functional parameter), the value of the P/R ratio was stable at approximately 1 under both control and irradiation conditions during the experimental period; no remarkable influence of gamma beam irradiation was recognized, and the behavior of DO was very stable (Fig. 7.33).

The structural parameter of the microcosm system was affected when the populations of certain microorganisms decreased or increased, but the functional parameter that assumed the P/R ratio by consecutive irradiation of the gamma beam did not come under influence. Mechanistically, even if a population increases or decreases, the metabolic activity of the population was not affected, and accordingly the possibility that the metabolic activity of an individual changed was suggested. For example, it is thought that the metabolic activity of an individual that survived rose when a population declined. Additionally, when the population of a certain organism increased or decreased, and the metabolic activity of the population changed, the metabolic activity of organisms with similar ecological functions changed accordingly, and the possibility that the metabolic activity of the population was not affected is considered. In other words, it is thought that when the population of an organism decreased, and its metabolic activity decreased, other organisms' populations rose using a surplus of resources. By either mechanism, the results obtained suggest the ability of the microcosm to be maintained functionally at the population or community level, even in the case in which the microcosm has been structurally affected (Fuma et al. 2010).

It is thought that the possibility that a real ecosystem will be bombed at equivalent dose rates to these experiments (i.e., 10 Gy/day and 23 Gy/day radiation) is extremely low. Serious atomic energy accidents do occur, and the inappropriate disposal of high-level atomic waste has been practiced, but most of the associated dose rates are less than the values used in this experiment. For example, in the Chernobyl nuclear power plant accident, which occurred in the former Soviet Union (present-day Ukraine) in 1986 and was said to be the worst nuclear disaster ever, the maximum dose rate that a fish received was only 0.03 Gy/day. Additionally, in the Mayak nuclear compound, also in the former Soviet Union (present-day Russia) and south of the Ural region, it is thought that an individual fish received a maximum dose of radiation equal to 0.6 Gy/day due to the inappropriate disposal of radioactive waste into the Techa River from 1950 to 1951, and a maximum of 0.1 Gy/day was provided in the Kyshtym accident due to malfunctioning of the cooling facilities, which occurred in 1957. The only example of a natural environment being bombed with dose rates greater than those examined in this experiment is due to the inappropriate disposal of large quantities of radioactive waste to Lake Karachay in the southern Ural Mountains from 1951 through 1952, when the dose rate was

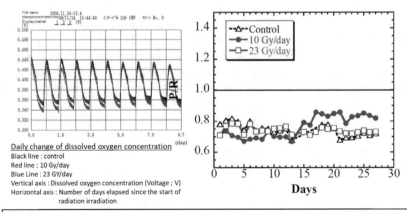

Daily change of dissolved oxygen concentration
Black line : control
Red line : 10 Gy/day
Blue Line : 23 GY/day
Vertical axis : Dissolved oxygen concentration (Voltage ; V)
Horizontal axis : Number of days elapsed since the start of radiation irradiation

Since the value of the control is smaller than 1, there are some points to improve in the measurement·calculation method. However, it seems that radiation does not significantly affect P/R.

It is difficult to think that it will exceed the dose rate of this experiment in the actual environment even if accident of nuclear facility is considered*. Therefore, it is unlikely that aquatic microbial communities will be greatly affected by radiation.

*The only exception: Dose rate of the former Soviet Union Karachai Lake, in which a large amount of radioactive waste was dumped in the 1950s, was up to 800Gy/day. By the way, the dose rate at the time of the Chernobyl nuclear accident is about 0.03Gy/day at the maximum.

Fig. 7.33 Assessment from P/R ratio of γ-ray (^{137}Cs)

estimated to reach 300–800 Gy/day, if it was assumed that the fish survived. Therefore, the radiation risk for serious damage to aquatic microbial ecosystems is expected to be low, even if the accidents at atomic energy facilities and the inappropriate disposal of radioactive waste are considered, because a change in population was observed with some microorganisms in a microcosm, without any concomitant influence on the P/R ratio of the microcosm, in the 23 Gy/day experiments (Fuma et al. 2012) (Fig. 7.33).

Using another aquatic microcosm, a more simplified system, the effects of chronic γ-irradiation were investigated in the microcosm consisting of flagellate alga, *Euglena gracilis* Z, as producers; the ciliate protozoan, *Tetrahymena thermophila* B, as consumers; and the bacterium, *Escherichia coli* DH5α, as decomposers. At a dose rate of 1.1 Gy/day, no effects were observed. At a dose rate of 5.1 Gy/day, the population of *Escherichia coli* showed a tendency to be lower than that of the control system. At dose rates of 9.7 Gy/day and 24.7 Gy/day, a population decrease was observed in *Escherichia coli*. *Euglena gracilis* and *Tetrahymena thermophila* died out after a temporary population decrease, and the abundance of *Tetrahymena thermophile* subsequently increased. It is likely that this temporary population increase was an indirect effect of interspecies interactions. The effect dose rates of γ-rays were compared with the effect concentrations of some metals using a radiochemoecological conceptual model and the effect index for the microcosm. Comparison of these community-level effects with environmental exposure data suggests that ionizing radiation, Gd, and Dy pose low risks to aquatic microbial communities, while Mn, Ni, and Cu pose considerable risks. The effects of chronic irradiation were smaller than those of acute irradiation, and an acute-to-chronic ratio was calculated to be 28 by dividing an acute dose by the chronic daily dose rate at which the effect index was 10%. This ratio would be useful for community-level extrapolation from acute to chronic radiation effects.

It is necessary to evaluate the combined effects of ionizing radiation and other toxic agents on ecosystems because ecosystems are exposed to various factors. The combined effects of γ-rays and acidification on an experimental model ecosystem (i.e., the microcosm) that mimicked aquatic microbial communities were investigated. Microcosms, consisting of *Euglena gracilis* Z as a producer, *Tetrahymena thermophila* B as a consumer, and *Escherichia coli* DH5α as a decomposer, were loaded by the following treatments: (1) irradiation with 100 Gy ^{60}Co γ-rays; (2) acidification of the culture medium to a pH = 4.0, with a mixture of 0.1 N of HNO_3 and 0.1 N of H_2SO4 (1:1, v/v), which mimicked acid rain; and (3) irradiation with 100 Gy γ-rays followed by the acidification of the culture medium (pH = 4.0). The γ-irradiation induced a temporary decrease in the cell density of *Escherichia coli* but did not affect the cell densities of the other species. The concentrations of chlorophyll *a* and ATP in the microcosm were not affected by γ-irradiation, and chlorophyll *a* concentrations in *Euglena gracilis* cells were also not affected. Acidification significantly decreased the cell density of *Tetrahymena thermophila*, slightly decreased the cell density of *Escherichia coli*, and slightly increased the cell density of *Euglena gracilis*. The concentrations of chlorophyll *a* and ATP in the microcosm were increased by acidification, although chlorophyll *a* concentrations in *Euglena gracilis* cells decreased. The combined exposure to γ-rays and acids temporarily decreased the cell density of *Escherichia coli*, significantly decreased the cell density of *Tetrahymena thermophila*, and slightly increased the cell density of *Euglena gracilis*. The concentrations of chlorophyll *a* and ATP in the microcosm were increased by this combined exposure, although chlorophyll *a* concentrations in *Euglena gracilis* cells decreased. It was therefore concluded that the combined exposure to γ-rays and acids had additive effects on cell densities, chlorophyll *a* and ATP concentrations in the microcosm, and chlorophyll *a* concentrations in *Euglena gracilis* cells (Fuma et al. 2010).

7.11.2   γ-Ray (^{60}Co)

^{60}Co γ-rays were used to irradiate the microcosm at various stages of biological succession and with various strengths, and the microcosm was transferred to a fresh culture medium after a certain period of time. In the case of irradiation during the initial stage or young stage, with various strengths, all microorganisms in the microcosm became extinct at 3000 R and the system collapsed; it did not recover by being transferred to a fresh culture medium. In contrast, the system did not collapse with 1000,000 R irradiation during the stable stage. In the transferred, fresh culture medium after 1 day since irradiation, there was an increased volume of biomass, which was indicated to be higher in 1000,000 R <100,000 R <10,000 R, and close to that of the control system. However, in the case of transferring after 22 days, the 1,000,000 R irradiated system showed a similar increase in the volume of biomass as observed in the control system. This indicates that damage from irradiation was recovered after 22 days. The fact the mature biota had high fecundity means that microorganisms survived under strong irradiation, and these surviving individuals recovered to increasing conditions again after 22 days.

The microorganisms in the microcosm N-system were not influenced by irradiation stronger than that just after the Fukushima nuclear power plant accident in Japan. This phenomenon makes sense considering that microorganisms first appeared on the early Earth, which was exposed to very strong irradiation. It is important to generate new findings for environmental protection and restoration that some microorganisms have the ability to resist the influence of strong irradiation.

7.12   Microbial Pesticides

These days, the residues and accumulation of chemical pesticides have become one of the most serious problems all over the world. At the same time, biological prevention such as the use of microbial pesticides, which include natural enemies, is revisited, and, in Europe and the USA, microbial pesticides have already been made practicable. They are obtained from organisms including plants, bacteria and other microbes, fungi, nematodes, etc. They are often important components of integrated pest management (IPM) programs and have received much practical attention as substitutes for synthetic chemical plant protection products (PPPs).

Microbial pesticides are one of many biological preservation methods, and their main components are bacteria, fungi, viruses, protozoans, and nematodes, which may or may not be. These microbial pesticides are considered safer for the environment than chemical pesticides because of their natural origins. Furthermore, not only natural microbial pesticides but also genetically engineered ones have been developed, and more capitalization has been done. Note, however, that how the microbial pesticide behaves in nature and what effects will happen have not been made clear. Therefore, it is very important to make clear the proliferation and decay of microbial pesticides in natural ecosystems and to obtain the basal information for environmental assessment for the field release of microbial pesticides.

7.12.1   Bacillus thuringiensis *subsp.* aizawai *KH*

In this section, we focus on the environmental impact assessment of microbial pesticides, which constitute the lowest prey species in the microcosmic food chain. *Bacillus thuringiensis* subsp. *aizawai* KH was added at 1, 10, and 100 times the abundance of indigenous bacteria in the microcosm, as a bacterial pesticide, and its influence on the structure and function of the ecosystem was examined. This strain has Sm^r and Rf^r as markers. *Bacillus thuringiensis* subsp. *aizawai* KH rapidly decreased at 1, 10, or 100 times inoculation concentrations in viable counts using the selective media method, following microcosm inoculation, and the abundance of microorganisms (in particular, its primary consumer, the protozoan ciliate, *Cyclidium glaucoma*) in the microcosm increased. *Bacillus thuringiensis* subsp. *aizawai* KH has shown to be a suitable food source for constituent heterotrophic groups such as protozoan ciliates, including *Cyclidium glaucoma*, *Philodina erythrophthalma*, and *Aeolosoma hemprichi*, in prey-predator interaction tests, which were considered to be dominated by these predatory actions. By comparison, in N_{30}, no significant difference was observed between the system with *Bacillus thuringiensis* subsp. *aizawai* KH added and the system without it (control system).

Bacillus thuringiensis subsp. *aizawai* KH, as a microbial pesticide, showed the same behavior as *Bacillus cereus* MC, one of the indigenous bacteria acting as a control. That is, these foreign bacteria decreased in the microcosm and did not strongly influence the indigenous microorganisms in the microcosm. As described in Fig. 7.4, the population density of *Bacillus thuringiensis* subsp. *aizawai* KH was 7.8×10^7 CFU/mL in the injected day (16th day) and decreased to 1.1×10^6 CFU/mL 2 days after (18th day), 2.2×10^4 CFU/mL 7 days after (23rd day), and 9.8×10^3 CFU/mL 14 days after (30th day). In a manner similar to the population density of *Bacillus thuringiensis* subsp. *aizawai* KH, the spores did not increase in abundance but survived at a fixed density of 5.8×10^2 N/mL after injection into the microcosm. This was because heterotrophs disliked the *Bacillus thuringiensis* spores as a food source, and vegetative cells produced spores without interruption. It was made clear that the predator-prey interaction between *Bacillus thuringiensis* and heterotrophs played a significant role in the proliferation and decay of microbial pesticides. After the injection of *Bacillus thuringiensis* and *Bacillus cereus*, a rapid increase in the number of protozoa, especially *Cyclidium glaucoma*, was observed. This suggested that the predator-prey interaction between foreign bacteria injected, such as microbial pesticide and protozoa, existed. This suggests that the proliferation and decay of microbial pesticides were greatly affected by the predation of heterotrophs, especially protozoa, which inhabited the microcosm system.

From an estimation of the succession pattern of microorganisms as the structural parameter, the protozoa *Cyclidium glaucoma* was strongly influenced by *Bacillus thuringiensis* subsp. *aizawai* KH, and this protozoa increased in its abundance under the *Bacillus thuringiensis* addition, and other microorganisms, such as the rotifers, *Lecane* sp. and *Philodina erythrophthalma*, also increased compared with the control microcosm. Thus, the influence of *Bacillus thuringiensis* was different in different microorganisms in coexisting culture conditions, such as the microcosm system. The water quality, that is, the pH, rose with the *Bacillus thuringiensis* addition, but the value of the pH was approximately 8.5 to 9.5 across all experiments. From an estimation of N_{30} using the *Bacillus thuringiensis* addition, all species of microorganisms increased in abundance. From an estimation of B_{16-30}, the same as in N_{30}, all microorganisms increased (Fig. 7.34).

Moreover, almost the same behavior was seen from the introduced *Bacillus thuringiensis* subsp. *aizawai* KH as bacterial microbial pesticide was observed in a

Fig. 7.34 Population
dynamics in *B.t. aizawai*
KH-added microcosm
N-system

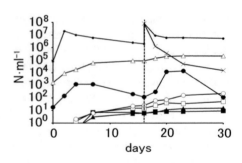

natural lake model ecosystem, that is, a naturally derived microcosm, which was
made from natural lake water including natural microorganisms and nutrients.

7.12.2 Bacillus thuringiensis *subsp.* kurstaki

The environmental impact risk assessment of microbial pesticides, which constitute
the lowest prey species in the microcosmic food chain, was studied. *Bacillus
thuringiensis* subsp. *kurstaki*, as a bacterial pesticide, was added at 1, 10, and
100 times the abundance of indigenous bacteria in the microcosm, and the influence
on the structure and function of the ecosystem was examined. *Bacillus thuringiensis*
subsp. *kurstaki* rapidly decreased at 1, 10, or 100 times the inoculation concentrations
in viable counts using the selective media method, following microcosm inoculation,
and the number of microorganisms (in particular, its primary consumer, the protozoan
ciliate *Cyclidium glaucoma*) in the microcosm increased. *Bacillus thuringiensis* subsp.
kurstaki has been shown to be a suitable food source for constituent heterotrophic
groups such as protozoan ciliates, including *Cyclidium glaucoma*, *Philodina
erythrophthalma*, and *Aeolosoma hemprichi* in prey-predator interaction tests, which
were considered to be dominated by these predatory actions. By comparison in N_{30}, no
significant difference was observed between the system with *Bacillus thuringiensis*
subsp. *kurstaki* added and the system without it (control system).

7.12.3 Pseudomonas fluorescens *IID5115*

Pseudomonas fluorescens IID5115, as a bacterial pesticide, was added at 1, 10, and
100 times the abundance of indigenous bacteria in the microcosm, and its influence
on the structure and function of the ecosystem was examined. *Pseudomonas
fluorescens* IID5115 rapidly decreased at 1-, 10-, or 100-times inoculation concentrations in viable counts using the selective media method after microcosm inoculation, and the abundance of microorganisms (in particular, its primary consumer,

the protozoan ciliate *Cyclidium glaucoma*) in the microcosm increased. *Pseudomonas fluorescens* IID5115 has been shown to be a suitable food source for constituent heterotrophic groups. By comparison, in N_{30}, no significant difference was observed between the system with *Pseudomonas fluorescens* IID5115 added and the system without it (control system).

Pseudomonas fluorescens IID5115, as a bacterial pesticide, was added in 10^7 N/mL; it decreased slightly after suddenly falling to 10^3 N/mL on the 30th day and disappeared finally. Meanwhile, the population density of the ciliate, *Cyclidium glaucoma*, increased from 10^2 N/mL to 10^4 N/mL. It was suggested that the predation pressure of microorganisms has a strong influence on the proliferation and decay of indigenous bacteria introduced into the ecosystem, and the results of the predator-prey interaction monoxenic culture test was lined.

7.12.4   Beauveria bassiana *F18-4B*

Beauveria bassiana F18-4B, as a fungal pesticide, was added at one and ten times the abundance of indigenous bacteria in the microcosm, and its influence on the structure and function of the ecosystem was examined. *Beauveria bassiana* F18-4B slowly decreased at 1- or 10-times inoculation concentrations in viable counts using the selective media method, after microcosm inoculation, and the number of microorganisms in the microcosm increased. *Beauveria bassiana* F18-4B has been shown to be a suitable food source for *Philodina erythrophthalma* and *Aeolosoma hemprichi*, in predator-prey interaction tests, which was considered to be dominated by these predatory actions. By comparison, in N_{30}, no significant difference was observed between the system with *Beauveria bassiana* F18-4B added and the system without it (control system). Moreover, nearly identical behavior of the introduced *Beauveria bassiana* F18-4B, as a fungal microbial pesticide, was observed in a natural lake model ecosystem, that is, a naturally derived microcosm, which was made from natural lake water, including natural microorganisms and nutrients. Irrespective of the kind of natural environmental water, all fungal pesticides decreased and disappeared in the system finally after it was added at 10^5 N/mL. The individual number of micro flagellates, such as *Chlamydomonas* sp. and *Monas* sp., in the fungal pesticide addition system increased from 10^2 N/mL to 10^3 N/mL; compared with the control system, the influence of the predation pressure by heterotrophs was confirmed. Microbial pesticides, both fungal and bacterial, exhibited a decreasing trend in the systems from which an organismal factor was removed through filtration processing, but the survivorship period was clearly lengthy, and the influence of the interaction between the microbes and the indigenous microorganisms was also confirmed. For indigenous microorganisms to become so varied in the eutrophied aquatic environment, a survivorship period of microbial pesticide tended to become short, so that the degree of water pollution was high, and a small amount of chemicals which caused the water contamination restrained the life and reproduction of the microbial pesticide. Organic matter also had a high degree of pollution, the influence of the interaction by which the added microbial pesticide acted as competition with a settled microorganism and decreased. Or it was supposed that the physicochemical-like factor, which is the water quality, as well as the interaction between the organisms in the behavior of microbial pesticides is closely related.

7.12.5   Verticillium lecanii *F126-12-3M*

Verticillium lecanii F126-12-3M, as a fungal pesticide, was added at one and ten times the abundance of indigenous bacteria in the microcosm, and its influence on the structure and function of the ecosystem was examined. *Verticillium lecanii* F126-12-3M slowly decreased at 1- or 10-times inoculation concentrations in viable counts using the selective media method after microcosm inoculation, and the abundance of small microorganisms in the microcosm increased. *Verticillium lecanii* F126-12-3M has been shown to be a suitable food source for only *Philodina erythrophthalma* and *Aeolosoma hemprichi* in the predator-prey interaction test, which was considered to be dominated by these predatory actions. By comparison, in N_{30}, no significant difference was observed between the system with *Verticillium lecanii* F126-12-3M added and the system without it (control system).

7.12.6   Metarhizium anisopliae *M7*

Metarhizium anisopliae M7, as a fungal pesticide, was added at one and ten times the abundance of indigenous bacteria in the microcosm, and its influence on the structure and function of the ecosystem was examined. *Metarhizium anisopliae* M7 slowly decreased at 1- or 10-times inoculation concentrations in viable counts using the selective media method after microcosm inoculation, and the abundance of microorganisms in the microcosm increased. *Metarhizium anisopliae* M7 has been shown to be a suitable food source for constituent heterotrophic groups such as metazoan *Philodina erythrophthalma* and *Aeolosoma hemprichi*. By comparison, in N_{30}, no significant difference was observed between the system with *Metarhizium anisopliae* M7 added and the system without it (control system).

7.13   Genetically Modified Bacteria

With the advancement of genetic engineering, the release of useful new microorganisms (e.g., genetically modified bacteria, genetically engineered microorganisms) to the environment has been investigated in various fields. From the viewpoint of environmental protection, their safe use should be confirmed before regular environmental release. It is essential to elucidate the behavior of GEMs in natural ecosystems such as rivers, lakes, and marshes and in artificial ecosystems including activated sludge and biofilms. Especially, interactions among microorganisms, such as prey-predator interactions, are significant (Sudo et al. 1990). Using a microcosm system for the estimation of the proliferation and decay of GEMs is a useful method, because the microcosm has very high reproducibility and reflects the natural ecosystem.

7.13.1   Escherichia coli *HB101/pBR325*

Escherichia coli is one species of bacillus of gram-negative rods and is classified as facultative anaerobic bacteria. These bacteria are one of the major species that are present not only in the environment but also in the digestive tract of warm-blooded animals (birds and mammals), especially humans. Their size is usually 0.4–0.7 μm along the short axis and 2.0–4.0 μm along the long axis, but the long axis can be shortened, and some cells are nearly spherical. It is a model organism representative of bacteria, and it is used as a material in various studies, such as in genetic engineering, and also for the production of chemical substances through gene incorporation. Here, we focused on the competitive relationship among four taxa of bacteria, which are the lowest organisms on the microcosmic food chain. *Escherichia coli* HB101/pBR325 was added at 1, 10, and 100 times the abundance of indigenous bacteria in the microcosm, and its influence on the structure and function of the ecosystem was examined. The pBR325, a 5.9-kbp-sized vector plasmid, which is widely used for gene manipulation, is a non-transmissible plasmid, coding Cm^r, Tc^r, and Ap^r and having Eco *RI*, Bam *HI*, Hind *III*, and Sal *I* sites on its nucleotide sequence.

Escherichia coli HB101/pBR325 rapidly decreased at 1-, 10-, or 100-times inoculation concentrations in viable counts using the selective media method following microcosm inoculation, and the abundance of microorganisms (in particular, its primary consumer, the protozoan ciliate *Cyclidium glaucoma*) in the microcosm increased. *Escherichia coli* HB101/pBR325 has been shown to be a suitable food source for constituent heterotrophic groups in the microcosm. By comparison, in N_{30}, no significant difference was observed between systems with *Escherichia coli* added and systems without it (control system). *Escherichia coli* HB101/pBR325, as GEMs, decreased in its abundance in the same manner as *Bacillus thuringiensis*. The population density of *Escherichia coli* was 3.6×10^7 CFU/mL on the injected day (16th day) and rapidly decreased to 2.3×10^5 CFU/mL 2 days after (18th day), 7.2×10^3 CFU/mL 7 days after (23rd day), and 6.8×10^3 CFU/mL 14 days after (30th day). The predation effect of the protozoan *Cyclidium glaucoma* was observed at the same time. This indicates that *Escherichia coli* HB101/pBR325 was a suitable food source for the indigenous heterotrophs in the microcosm.

From an estimation of the succession pattern of microorganisms, *Cyclidium glaucoma* was strongly influenced by *Escherichia coli* HB101/pBR325, especially since this protozoan's density increased compared to other taxa, such as rotifer *Philodina erythrophthalma*, which slightly increased in abundance under these conditions, and compared with the control microcosm. From an estimation of N_{30}

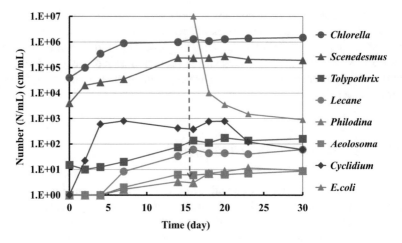

Fig. 7.35 Prosperity and decay of *E.coli* HB101/pBR325 in microcosm N-system

with *Bacillus cereus* as one of the indigenous bacteria additions, all species of microorganisms increased in abundance. From an estimation of B_{16-30}, all microorganisms increased similarly to N_{30}. In the case of the *Escherichia coli* addition, the trends of N_{30} and B_{16-30} were almost the same as in the case of the *Bacillus cereus* addition. In addition, from the attenuation patterns of other foreign bacteria as GEMs, the patterns were divided into three types, that is, (a) rapidly decreasing, (b) slowly decreasing, and (c) slightly decreasing. *Bacillus thuringiensis* was classified as type b, and *Escherichia coli* was type a or b. Thus, the fate of foreign bacteria (alien species) can be assessed from the viewpoint of their attenuation pattern using a microcosm test (Fig. 7.35).

In the P/R ratio, which is the ratio of the production and consumption of dissolved oxygen, the consumption activity of dissolved oxygen (DO) increased as the amount of added *Escherichia coli* HB101/pBR325 increased, but eventually the non-additive system (control system) converged on the same level, and the P/R ratio was also stabilized at about 1. For this reason, the introduction of *Escherichia coli* HB101/pBR325 was judged as having no influence on the ecosystem resulting from the variation in the structural type of the structural parameter and the decline of the DO value of the functional parameter. Even if the lowest species of the food chain invades a microcosm, it was evaluated that there was no major impact on the ecosystem. When introducing the lowest species on the food chain (*Escherichia coli* HB101/pBR325) into the microcosm, both structural and functional parameters converge to the same extent as in the control system (non-additive system), and the existing ecosystem is weakly affected.

7.13.2   Escherichia coli *S17-1/pSUP104*

Escherichia coli S17-1/pSUP104 was added at 1, 10, and 100 times the abundance of indigenous bacteria in the microcosm, and its influence on the structure and function of the ecosystem was examined. The pSUP104 is a mobilized transmissible plasmid through the chromosome.

   Escherichia coli S17-1/pSUP104 rapidly decreased at 1-, 10-, or 100-times inoculation concentrations in viable counts using the selective media method, following microcosm inoculation, and the abundance of microorganisms (in particular, its primary consumer, *Cyclidium glaucoma*) in the microcosm increased. *Escherichia coli* S17-1/pSUP104 has been shown to be a suitable food source for constituent heterotrophs in the microcosm. By comparison, in N_{30}, no significant difference was observed between the system with *Escherichia coli* added and the system without it (control system). Moreover, a nearly identical behavior of introduced *Escherichia coli* S17-1/pSUP104, as a GEM, was observed in a natural lake model ecosystem, that is, a naturally derived microcosm, which was made from natural lake water including natural microorganisms and nutrients.

7.13.3   Escherichia coli *C600/RP4*

Escherichia coli C600/RP4 was added at 1, 10, and 100 times the abundance of indigenous bacteria in the microcosm, and its influence on the structure and function of the ecosystem was examined. The RP-4, 56-kbp-sized vector plasmid, which is widely used for gene manipulation, is a self-transmissible plasmid, coding Ap^r, Tc^r, and Km^r and having an Eco *RI* site on its nucleotide sequence.

   Escherichia coli C600/RP4 rapidly decreased at 1-, 10-, or 100-times inoculation concentrations in viable counts using the selective media method, following microcosm inoculation, and the abundance of microorganisms (in particular, its primary consumer, *Cyclidium glaucoma*) in the microcosm increased. *Escherichia coli* C600/RP4 has been shown to be a suitable food source for constituent heterotrophs in the microcosm. By comparison, in N_{30}, no significant difference was observed between the system with *Escherichia coli* added and the system without it (control system).

7.13.4   Escherichia coli *S17-1/pCRO1*

Escherichia coli S17-1/pCRO1 was added at 1, 10, and 100 times the abundance of indigenous bacteria in the microcosm, and its influence on the structure and function of the ecosystem was examined. The pCRO1, with a 1580 bp fragment size, is a mobilized transmissible plasmid through the chromosome, coding Cr^r on its nucleotide sequence. This plasmid was prepared by insertion between the ECO *RI* and Hind *III* sites of pBR322.

Escherichia coli S17-1/pCRO1 rapidly decreased at 1-, 10-, or 100-times inoculation concentrations in viable counts using the selective media method, following microcosm inoculation, and the abundance of microorganisms (in particular, its primary consumer, the protozoan ciliate *Cyclidium glaucoma*) in the microcosm increased. *Escherichia coli* S17-1/pCRO1 has been shown to be a suitable food source for constituent heterotrophs in the microcosm. By comparison, in N_{30}, no significant difference was observed between the system with *Escherichia coli* added and the system without it (control system). Moreover, a nearly identical behavior of the introduced *Escherichia coli* S17-1/pCRO1, as a GEM, was observed in a natural lake model ecosystem, that is, a naturally derived microcosm, which was made from natural lake water including natural microorganisms and nutrients.

7.13.5   Pseudomonas aeruginosa *PAO1/pCRO1*

Pseudomonas aeruginosa PAO1/pCRO1 was added at 1, 10, and 100 times the abundance of indigenous bacteria in the microcosm, and its influence on the structure and function of the ecosystem was examined. The pCRO1, with a 1580 bp fragment size, is a mobilized transmissible plasmid through the chromosome, coding Cr^r on its nucleotide sequence. This plasmid was prepared by insertion between the ECO *RI* and Hind *III* sites of pBR322.

Pseudomonas aeruginosa PAO1/pCRO1 slowly decreased at 1-, 10-, or 100-times inoculation concentrations in viable counts using the selective media method following microcosm inoculation, and the abundance of microorganisms (in particular, its primary consumer, the protozoan ciliate *Cyclidium glaucoma*) in the microcosm slowly increased. *Pseudomonas aeruginosa* PAO1/pCRO1 has been shown to be a suitable food source for constituent heterotrophs in the microcosm. By comparison, in N_{30}, no significant difference was observed between the system with *Pseudomonas aeruginosa* added and the system without it (control system). Moreover, a nearly identical behavior of the introduced *Pseudomonas aeruginosa* PAO1/ pCRO1, as a GEM, was observed in a natural lake model ecosystem, that is, a naturally derived microcosm, which was made from natural lake water including natural microorganisms and nutrients.

7.13.6   Pseudomonas putida *PpY101/pSR134*

Pseudomonas putida PpY101/pSR134 was added at 1, 10, and 100 times the abundance of indigenous bacteria in the microcosm, and its influence on the structure and function of the ecosystem was examined. The pSR134 codes Hg^r on its nucleotide sequence. *Pseudomonas putida* PpY101/pSR134 slowly decreased at 1-, 10-, or

100-times inoculation concentrations in viable counts using the selective media method, following microcosm inoculation, and the abundance of microorganisms (in particular, its primary consumer, the protozoan ciliate *Cyclidium glaucoma*) in the microcosm slowly increased. *Pseudomonas putida* PpY101/pSR134 has been shown to be a suitable food source for constituent heterotrophs in the microcosm. By comparison, in N_{30}, no significant difference was observed between the system with *Pseudomonas putida* added and the system without it (control system).

7.13.7 Pseudomonas putida *MC/pBR325*

Pseudomonas putida MC/pBR325 was added at 1, 10, and 100 times the abundance of indigenous bacteria in the microcosm, and its influence on the structure and function of the ecosystem was examined. *Pseudomonas putida* MC is one of the indigenous bacteria in the microcosm N-system. The pBR325, a 5.9-kbp-sized vector plasmid, which is widely used for gene manipulation, is a non-transmissible plasmid, coding Cm^r, Tc^r, and Ap^r and having Eco *RI*, Bam *HI*, Hind *III*, and Sal *I* sites on its nucleotide sequence. *Pseudomonas putida* MC/pBR325 slowly decreased at 1-, 10-, or 100-times inoculation concentrations in viable counts using the selective media method after microcosm inoculation, and the abundance of microorganisms (in particular, its primary consumer, the protozoan ciliate *Cyclidium glaucoma*) in the microcosm slowly increased. Pseudomonas putida MC/pBR325 has been shown to be a suitable food source for constituent heterotrophs in the microcosm. By comparison, in N_{30}, no significant difference was observed between the system with *Pseudomonas putida* added and the system without it (control system).

7.13.8 Pseudomonas paucimobilis *BHC+*

Pseudomonas paucimobilis BHC+ was added at 1, 10, and 100 times the abundance of indigenous bacteria in the microcosm, and its influence on the structure and function of the ecosystem was examined. *Pseudomonas paucimobilis* is a BHC+-degrading bacterium. *Pseudomonas paucimobilis* BHC+ slowly decreased at 1-, 10-, or 100-times inoculation concentrations in viable counts using the selective media method after microcosm inoculation, and the abundance of microorganisms (in particular, its primary consumer, protozoan ciliate *Cyclidium glaucoma*) in the microcosm slowly increased. *Pseudomonas paucimobilis* BHC+ has been shown to be a suitable food source for constituent heterotrophs in the microcosm. By comparison, in N_{30}, no significant difference was observed between the system with *Pseudomonas paucimobilis* added and the system without it (control system).

7.14 Biomanipulation

Changing the fish population of bodies of water as a part of watershed management can facilitate desirable changes in aquatic ecosystems suffering from eutrophication, which is characterized by phytoplankton dominance, thus aiding ecosystem restoration, an application of restoration ecology. In ponds or lakes, alternative stable conditions (i.e., one with high algal populations, little other plant life, and turbid water) and another with low algae populations, a diverse plant population, and clear water may exist. In addition, to prevent excess nutrients such as phosphorus and nitrates, the removal of certain fish species adapted to turbid water may facilitate change from one steady state to the other, through the application of dynamical systems theory. Fish species may be removed by means of poisoning, harvesting, or the introduction of predatory species. Since a different fish community will result from these processes, it will affect recreational and commercial fishermen whose cooperation is important.

Biomanipulation technology attracts attention as a method for improving the quality of the water. Biomanipulation is a method used to introduce a creature, but this method poses a danger by collapsing existing ecosystems. Therefore, the environmental assessment needed to perform biomanipulation is important. The experiments below were carried out to investigate the basic ecosystem impact statement with respect to the water quality improvement from the biomanipulation. This study performed the basic environmental assessment, which focused its attention on ecosystem function as the P/R ratio using a microcosm. The risk assessment of top-down control and bottom-up control as a biomanipulation method on an aquatic ecosystem was conducted using a flask-sized microcosm system and its production/respiration ratio and succession of microbial biota, in comparison with a control system.

7.14.1 Cyclidium glaucoma (Primary Consumer in the Microcosm N-System)

In this experiment, the microcosm was loaded with *Cyclidium glaucoma*, which is a primary predator in this gnotobiotic microcosm. For on-site biomanipulation, either the top predator or primary predator was introduced into a eutrophied ecosystem to ablate the irregular phytoplankton growth, such as in an algal bloom. As the structural parameter, the population of *Cyclidium glaucoma* decreased after the 20th day and converged equivalently with the control on the 30th day (Fig. 7.36). As for both microcosms, all microorganisms did not disappear from the system, and it was thought that the predator-prey interactions between the microbe and its prey functioned. As the functional parameter, the 10-times loaded system showed the same behavior as the control system. However, the amplitude (activity) of the 10-times loaded system is larger than that of the control system. The *Cyclidium glaucoma*-loaded system keeps its function as an ecosystem. However, this system caused a structural change. As the functional parameter, any P/R ratio of a loaded system becomes approximately

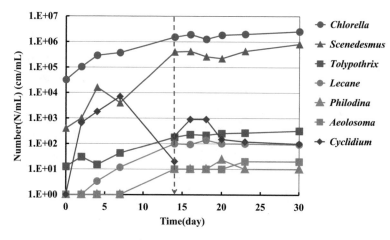

Fig. 7.36 Population dynamics in ten times of *Cyclidium*-added microcosm N-system

1. However, the activity of the system increased with the introduction of these loadings. As the structural parameter, the introduction of *Cyclidium glaucoma*, the primary predator, produces an activity change to the 10-times load, and it did not lead to the collapse of the ecosystem function.

7.14.2   Lecane *sp. (Primary Consumer in the Microcosm N-System)*

Lecane sp., one of the primary predators in the microcosm, was introduced into the microcosm. As the structural parameter, the population of *Lecane* sp. did not show significant change, but *Cyclidium glaucoma*, the primary consumer, increased to ten times that of the control and decreased after the 20th day. This was because the indigenous bacteria (decomposer and prey) took the metabolites of *Lecane* sp. and increased and bacteria-feeding microorganisms grew. However, because the specific growth rate (μ) of *Cyclidium glaucoma* (2.8/day) was larger than that of *Lecane* sp. (0.31/day), *Cyclidium glaucoma* increased. As the functional parameter, any P/R ratio of a loaded system becomes approximately 1. However, the activity of the system increased with the introduction of these loadings. As the structural parameter, the introduction of *Lecane* sp. produces an activity change to the 10-times load, and it did not lead to the collapse of the ecosystem function.

7.14.3   Philodina erythrophthalma *(Top Predator in the Microcosm N-System)*

Philodina erythrophthalma (one of the top predators in microcosm) was introduced into the microcosm as the introduced species for top-down biomanipulation. As a

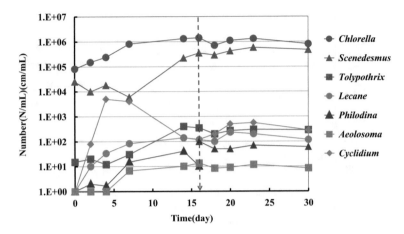

Fig. 7.37 Population dynamics in ten times of *Philodina*-added microcosm N-system

result of the mass addition of *Philodina erythrophthalma*, the color of the culture medium changed suddenly from light green to brown, and a brown solid accumulated at the bottom of the flask. Under microscopic observation, there were recognizable decolorized dead cells of *Chlorella* sp. Everywhere, and many were observed as crumbling. This is due to the excretion of much *Philodina erythrophthalma* which predated on *Chlorella* sp. At this point, both *Philodina erythrophthalma* and *Chlorella* sp. became extinct, and only bacteria and *Tolypothrix* sp. survived. As a result, in the flask with *Philodina erythrophthalma* introduced as the top predator species, *Chlorella* sp. decreased its population density by 45%, but *Scenedesmus quadricauda* did not. Although both *Aeolosoma hemprichi* and *Philodina erythrophthalma* are top predators in the microcosm N-system, *Aeolosoma hemprichi* is an aggregate feeder, while *Philodina erythrophthalma* is a bacterial feeder. Therefore, the effect of the top predator's predation type contributes significantly to the ecosystem of the microcosm. The top predator may keep down a proper population of microalga, and, in the case of top predator introduction, a possibility that could ameliorate algal blooms was demonstrated.

As a structural parameter, in either case, it is thought that the predator-prey interaction was constructed equally in the control system, without being removed from a system. The population of *Philodina erythrophthalma* slightly decreased in either 1-times quantity or 10-times quantity additions to a system by the 30th day while repeating the increase and decrease after addition together (Fig. 7.37). Additionally, for the decrease in chlorophyceans, the food source of rotifers, following the introduction of *Philodina erythrophthalma*, *Chlorella* sp. in the 10-times quantity addition system decreased by 45%, but *Scenedesmus quadricauda* did not decrease. Conversely, it is thought that the bacterial feeders *Cyclidium glaucoma* and *Lecane* sp. increased their populations mainly by having been dismantled as a metabolism by-product after *Cyclidium glaucoma* is added at ten times the quantity in the system, and *Cyclidium glaucoma* and *Lecane* sp. increase in the 1-times quantity addition system; *Chlorella* sp. was preyed on by *Philodina erythrophthalma*. It is thought that a 1-times quantity addition of *Philodina erythrophthalma* is enough to control a population of the chlorophyceans because there is no difference seen in the decrease in *Chlorella* sp. between the 1-times quantity and 10-times quantity from these additions and from the comparison with B_{16-30}.

As a functional parameter, from the comparison of DO and the P/R ratio, because the amplitude could be confirmed in both systems, production and respiration were shown to function in accordance with the light and dark period in the system. In addition, an activity increase was observed with a quantity of addition, same as in other addition systems. The P/R ratio converged at P/R = 1, and it followed that the ecosystem function did not come to change with the 10-times quantity addition of *Philodina erythrophthalma*.

7.14.4 Aeolosoma hemprichi *(Top Predator in the Microcosm N-System)*

The microcosm was loaded with *Aeolosoma hemprichi*, which is the top predator in this gnotobiotic microcosm. For on-site biomanipulation, the top predator and/or primary predator were introduced into the eutrophied ecosystem to reduce irregular phytoplankton growth, such as an algal bloom. As the structural parameter, the green algae *Chlorella* sp. and *Scenedesmus quadricauda* largely decreased after the addition of *Aeolosoma hemprichi*, in a 10-times load system. The decrease of *Chlorella* sp. and *Scenedesmus quadricauda* depends on the predation of *Aeolosoma hemprichi*. The 10-times load system was effective for large decreases of the phytoplankton. *Cyclidium glaucoma*, the primary predator, was increased to about five times, but *Lecane* sp. just decreased in the community with three kinds of zooplankton as predators. Because the growth rate of the bacteria was the highest compared to the other constituent microorganisms, *Cyclidium glaucoma* preyed on the increased bacteria, and it was thought that *Cyclidium glaucoma* populations increased. As the functional parameter, the amplitude of the 10-times loaded system is larger than the 1-times loaded system. In a natural ecosystem, creatures with an ecological position equivalent to that of *Aeolosoma hemprichi* and *Cyclidium glaucoma* could potentially be introduced in on-site biomanipulation. It was shown that the introduced ecosystem recovers to a state similar to before the introduction by keeping down microorganisms and the detritus when indigenous phytoplankton decreased within a system.

When *Aeolosoma hemprichi*, the top predator, was introduced, the green algae and cyanophyceans did not decrease in the 1-times loaded system under its predation. However, a population of the green algae was reduced to half just after the addition in the 10-times load. Therefore, it was suggested that the introduction of the top predator (top-down) had an influence on the ecosystem structure (Fig. 7.38).

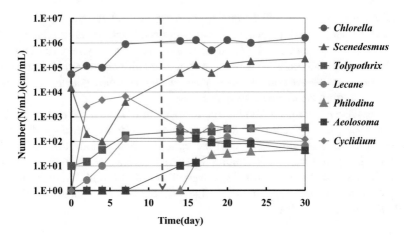

Fig. 7.38 Population dynamics in ten times of *Aeolosoma*-added microcosm N-system

However, the introduced ecosystem recovers to a state similar to that before the introduction by keeping down the introduced microorganisms and the detritus when the indigenous phytoplankton decreases within a system. It is necessary to remove the introduced microorganisms and detritus out of the system through the on-site biomanipulation. The microcosm was also effective as a tool for the environmental risk assessments, which targeted ecosystem function.

7.14.5 Moina macrocopa *(Alien Species)*

Moina macrocopa is filter-feeding minute crustacean belonging to *Daphnia*, with a physical size of 0.6–1.2 mm, and it is usually parthenogenetic (only in females), but males appear with the deterioration of the environment, due to water pollution and other stresses leading to the formation of resistant eggs. They are also useful as live baits for ornamental fishes and toxicity testing of chemicals, and they have gained attention as a biomanipulation-introduced species, among other applications. In this section, focusing on the competitive relationship with *Aeolosoma hemprichi*, which is the top predator in the microcosm, *Moina macrocopa* is added at one, five, and ten times the introduction amount of *Aeolosoma hemprichi* individuals, and the structure of the ecosystem and its effect on function are studied.

In any of the addition systems, the producers *Chlorella* sp. and *Tolypothrix* sp. decreased immediately after the introduction of *Moina macrocopa* (Fig. 7.39). The cause was thought to be that predation pressure increased from the *Moina macrocopa* invasion and *Tolypothrix* sp. could not grow sufficiently long due to amputation by the swimming behavior of *Moina macrocopa*. In predatory taxa, *Lecane* sp. and *Cyclidium glaucoma* showed a decreasing tendency, and, in the 10-times addition system, extinction was confirmed on the 18th day immediately

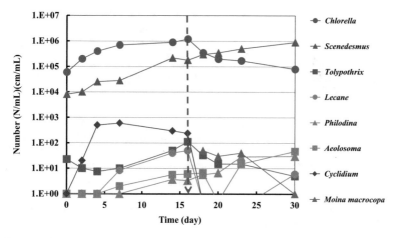

Fig. 7.39 Population dynamics of *Moina*-added microcosm N-system

after addition. This is thought to be indicative that the competitive relationship of consumers in the microcosm became severe due to the predation pressure of *Moina macrocopa*, and the differences in superiority or inferiority of predation ability were exhibited. In addition, the *Moina macrocopa* added as a foreign organism in all additive systems converged on the 30th day, by the time of measurement. In the 10-times addition system, assuming the invasion of a larger amount of alien species, the number of individuals of *Aeolosoma hemprichi* increased dramatically on the 30th day. This is because the existence of metabolites generated by the molting remnants of *Moina macrocopa* was a physical obstacle to filter-feeding zooplankton's predation behavior, and only *Aeolosoma hemprichi*, the detritus feeder, which has an opening-based feeding intake, advantageously predated green algae *Chlorella* sp.

From the P/R ratio, it can be seen that the consumption activity of dissolved oxygen (DO) increases as the added amount of *Moina macrocopa* increases, immediately after the start of continuous measurement in the 10-times addition system (Fig. 7.40). After the DO value attenuated for about 70 h, it stabilized. For this reason, introduction of the 10-times amount determined that the ecosystem was affected by the variation in the structural type of the structural parameter and the attenuation of the DO value of the functional parameter. With the top level of the food chain (highest predator) invaded in the microcosm, the maximum no-effect concentration was estimated to be between five and ten times that of the existing top-level predator. Furthermore, since the *Lecane* sp. is drastically decreased in the 5-times addition system from the structural parameter, the m-NOEC (microcosm maximum effect-free concentration) is considered to be at a relatively low concentration in the 5- to 10-times systems. In the case where a top-level predator (*Moina macrocopa*) was introduced into the microcosm, two predators were confirmed to go extinct in the 10-times addition system. There was a risk that the existing top-level predators would increase due to the invasion of alien species that could become higher predators.

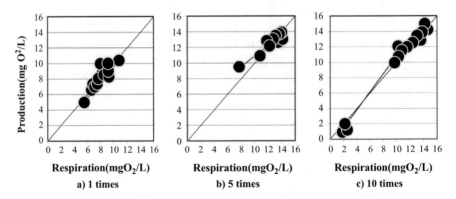

Fig. 7.40 Time course of P/R ratio in *Moina macrocopa*-added microcosm N-system. (**a**) One times. (**b**) Five times. (**c**) Ten times

7.14.6 Chlorella *sp. (Producer)*

Chlorella sp., one of the producers in the microcosm, was introduced into the microcosm as the introduced species for bottom-up biomanipulation. As structural parameters, no species became extinct within a culture period. The population of *Chlorella* sp. decreased in the 1-times quantity addition system from the time of addition to the 23rd day and increased afterward on the 30th day. On the other hand, the population of *Chlorella* sp. in the 10-times quantity addition system continued decreasing slightly from the time of addition to the 30th day and decreased approximately 78% more from the time of addition in the 30th day; this remarkable decrease in the population was confirmed. In addition, on the 30th day, the population of *Lecane* sp., *Philodina erythrophthalma*, and *Aeolosoma hemprichi*, which were the predators, were shown to decrease as *Chlorella* sp. was added. Whereas, the populations of *Cyclidium glaucoma* increased in comparison with the control system, and it followed that the quantity of existence increased. It is thought that the feeding niche of *Cyclidium glaucoma* broadened by the decrease in the population of three species above it in the food chain, which shared prey with *Cyclidium glaucoma*. The producers compete through the acquisition of the nutrient salts in the microcosm, which can also conjugate. Therefore, it is thought that the population increase in *Tolypothrix* sp. occurred because of the surplus nutrient salt, which was produced by a population decrease in *Scenedesmus quadricauda* and *Chlorella* sp.

As a functional parameter, in the case of *Chlorella* sp. in the 1-times quantity addition system, activity increased in comparison with the control system. This is regarded as an effect through the improvement of the photosynthetic efficiency from the increase in population. However, the DO just after the addition was in a state of high activity in the 10-times addition system, but the amplitude of the wave pattern decreased with the progression of the time, and an activity drop was shown. As for

10-times quantity addition system, the light inside of the flask is not at full intensity, and it is thought that the photosynthesis efficiency is low and counters the population growth of phytoplankton. This is a photosynthesis inhibition phenomenon, similar to algae in the natural ecosystem, and contributes to the extinction of fish from poor oxygenation in lakes and marshes. It is thought that the decrease in the population of *Chlorella* sp., which is a chlorophycean, is caused by a drop in their photosynthesis efficiency. However, it becomes a P/R ratio = 1 within each system, and it is thought that the ecosystem function in a 10-times addition system is low, but the system is maintained. From this, the 10-times quantity addition system was able to reproduce the decrease in activity from algae in a natural aquatic ecosystem, but it is thought that there is not recognized any influence such as the ecosystem collapses.

7.14.7   Scenedesmus quadricauda *(Producer)*

Scenedesmus quadricauda, one of the producers in the microcosm, was introduced into the microcosm as the introduced species for bottom-up biomanipulation. As a structural parameter, no species became extinct within a culture period. The population of *Scenedesmus quadricauda* decreased in the 1-times quantity addition system from the time of addition to the 23rd day, and it increased afterward on the 30th day. In contrast, the population of *Scenedesmus quadricauda* in the 10-times quantity addition system continued decreasing slightly from the time of addition to the 30th day.

As a functional parameter, in the case of *Scenedesmus quadricauda* in a 1-times quantity addition system, activity increased compared with the control system. This is regarded as an effect through the improvement of the photosynthetic efficiency from the increase in population. However, the DO just after the addition was in state of high in activity in the 10-times addition system, but the amplitude of the wave pattern decreased as time progressed, and an activity drop was shown. However, it becomes a P/R ratio = 1 within each system, and it is thought that the ecosystem function that a 10-times quantity addition system has is low, but the system is maintained. From this, the 10-times quantity addition system was able to reproduce the decrease in activity from algae in a natural aquatic ecosystem, but it is thought that there is no influence.

7.14.8   Tolypothrix *sp. (Producer)*

Tolypothrix sp., one of the producers in the microcosm, was introduced into the microcosm as the introduced species for bottom-up biomanipulation. As a structural parameter, each constituent organism class in the 1-times quantity addition system remained in the system. *Tolypothrix* sp. maintained a population in the 10-times quantity addition system until the 30th day, but *Cyclidium glaucoma* was not observed after the 18th day. *Aeolosoma hemprichi, Philodina erythrophthalma,* and *Cyclidium glaucoma* multiplied in the 1-times quantity addition system just after addition, compared to N_{30}, but *Cyclidium glaucoma* perished in the 10-times quantity addition system, and *Aeolosoma hemprichi*, *Philodina erythrophthalma*, and *Lecane* sp. remarkably increased from 4 times to 12 times in abundance. It is thought that the factor leading to the population increase in *Aeolosoma hemprichi*, *Philodina erythrophthalma*, and *Lecane* sp. was that it became easier to prey on microorganisms when *Cyclidium glaucoma* perished. Furthermore, it is thought that three kinds of predators multiplied in form to fill the niche of *Cyclidium glaucoma* in the microcosm as a result of the 10-times quantity added of *Tolypothrix* sp. because they have the characteristic of easily becoming an algal floc since *Tolypothrix* sp. is a *conferva*, and *Aeolosoma hemprichi* and *Philodina erythrophthalma* multiply in a cohesion body as a living space. In addition, it was thought that the extinction of *Cyclidium glaucoma* was caused by the fact that an algal floc of *Tolypothrix* sp. inhibited swimming. From the comparison of B_{16-30}, because the abundance of the predators, except *Cyclidium glaucoma*, increased with an increase in the quantity of addition, it was revealed that the addition of *Tolypothrix* sp. had an influence on the predator.

As a functional parameter, from the time course of DO and P/R ratio, it appears that the amplitude (activity) increases with an increase in the quantity of addition. *Tolypothrix* sp., a blue-green algae, photosynthesizes in the microcosm, but, when comparing the 10-times quantity addition system with the 1-times quantity addition system, the amplitude (activity) of the 10-times quantity addition was approximately two times that of the 1-times quantity addition, and it was shown that the addition quantity and activity did not necessarily agree. In addition, it is thought that the transition of the system occurred regardless of an active increase, although *Cyclidium glaucoma* perished in the 10-times quantity addition system, as for the P/R ratio because it is stable in P/R $= 1$.

7.14.9 **Bacillus cereus *MC (Decomposer)***

The competitive relationship among the four species of bacteria that are the lowest species in the microcosmic food chain was studied. *Bacillus cereus* MC, one of the indigenous bacteria in the microcosm, was added at 1, 10, and 100 times the abundance of indigenous bacteria in the microcosm, and its influence on the structure and function of the ecosystem was examined. *Bacillus cereus* MC slowly decreased at 1-, 10-, or 100-times inoculation concentrations in viable counts using the selective media method after microcosm inoculation, and the abundance of microorganisms (in particular, its primary consumer, the protozoan ciliate *Cyclidium glaucoma*) in the microcosm slowly increased. *Bacillus cereus* MC has been shown to be a suitable food source for microcosm heterotrophs. By comparison, in N_{30}, no significant difference was observed between the system with *Bacillus cereus* added and the system without it (control system).

7.14.10 Pseudomonas putida *MC (Decomposer)*

Pseudomonas putida MC, one of the indigenous bacteria in the microcosm, was added at 1, 10, and 100 times the abundance of indigenous bacteria in the microcosm, and its influence on the structure and function of the ecosystem was examined. *Pseudomonas putida* MC slowly decreased at 1-, 10-, or 100-times inoculation concentrations in viable counts using the selective media method after microcosm inoculation, and the abundance of microorganisms (in particular, its primary consumer, the protozoan ciliate *Cyclidium glaucoma*) in the microcosm slowly increased. *Pseudomonas putida* MC has been shown to be a suitable food source for microcosm heterotrophs. By comparison, in N_{30}, no significant difference was observed between the system with *Pseudomonas putida* added and the system without it (control system).

7.14.11 *Bacterial Mixed Culture*

A mixed culture of four species of indigenous bacteria in the microcosm, *Bacillus cereus*, *Pseudomonas putida*, *Acinetobacter* sp., and coryneform bacteria, was added at 1, 10, and 100 times the abundance of indigenous bacteria in the microcosm, and its influence on the structure and function of the ecosystem was examined. As a result, the total number of bacteria CFU slowly decreased at 1-, 10-, or 100-times inoculation concentrations in viable counts using the selective media method after microcosm inoculation, and the abundance of microorganisms (in particular, its primary consumer, the protozoan ciliate *Cyclidium glaucoma*) in the microcosm slowly increased. The bacterial mixed culture has been shown to be a suitable food source for microcosm heterotrophs. By comparison, in N_{30}, no significant difference was observed between the system with the bacterial mixed culture added and the system without it (control system).

7.14.12 Pseudomonas putida *prS2000 (Alien Species)*

Pseudomonas putida prS2000 as alien species was added at 1, 10, and 100 times the abundance of indigenous bacteria in the microcosm, and its influence on the structure and function of the ecosystem was examined. *Pseudomonas putida* prS2000 slowly decreased at 1-, 10-, or 100-times inoculation concentrations in viable counts using the selective media method after microcosm inoculation, and the abundance of microorganisms (in particular, its primary consumer, the protozoan ciliate *Cyclidium glaucoma*) in the microcosm slowly increased. *Pseudomonas putida* prS2000 has been shown to be a suitable food source for microcosm heterotrophs. By comparison, in N_{30}, no significant difference was observed between the system with *Pseudomonas putida* added and the system without it (control system).

7.14.13   Plasmid pBR325(Gene Biomanipulation)

As a form of bio-augmentation, *Escherichia coli* HB101/pBR325 decreased immediately after being added in the microcosm by counting CFU on the selective
medium plate, and, at the same time, the bacteriophagous ciliate, *Cyclidium glaucoma*, the primary predator, increased. It was indicated that the predator-prey
interaction between the introduced bacteria and the primary predatory protozoa
was influenced strongly. Furthermore, bacteriophagous metazoans, the rotifers
Lecane sp. and *Philodina erythrophthalma*, and the oligochaete, *Aeolosoma
hemprichi*, also increased slowly. From these outcomes, the proliferation and
decay of the introduced *Escherichia coli* HB101/pBR325 were considered to be
controlled by the predation of indigenous microorganisms.

As gene biomanipulation, in the case of the direct addition of the pBR325 plasmid
alone, there were no observed increases in heterotrophs in the microcosm, but the
CFU on the selective medium plate increased. This means the introduced non-
transmissible plasmid pBR325 was taken up by indigenous bacteria and phenotyp-
ically expressed its gene information, i.e., the non-transmissible plasmid can remain
by horizontal transferring between introduced bacteria (host strain) and indigenous
bacteria under the biological interactions in the microcosm (Fig. 7.41). From these
outcomes, *Escherichia coli* HB101/pBR325, the introduced bacteria, decreased
immediately through the control of the heterotroph's predation effect, but the
non-transmissible plasmid pBR325 can remain through bacterial horizontal transferring. Moreover, nearly the same behavior of the introduced *Escherichia coli*

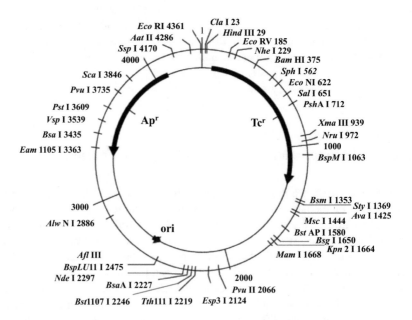

Fig. 7.41 Gene map of plasmid pBR325

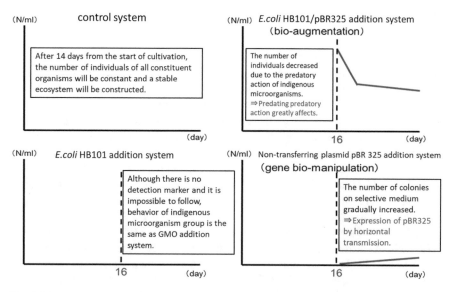

Fig. 7.42   Gene expression of non-transmissible plasmid pBR325 in microcosm N-system

HB101/pBR325 as a GEM was observed in a natural lake model ecosystem, that is, a naturally derived microcosm, which was made from natural lake water, including natural microorganisms and nutrients (Fig. 7.42).

From the asymptote, the microcosm N-system was shown to have the full stability of the system with respect to the population density of the original constituent microbial aspects, even if they are inoculated as constituent microbes from the microcosm N-system. In contrast, when it is inoculated with a foreign microbe, which was not a constituent microbe, it was shown that the influence on a system from highly advanced consumers was greater than in that from a lower-level prey introduction. It was also shown that the ecosystem impact statement, which includes the microbial interactions such as prey-predator interaction is able to be evaluated.

7.15   Climate Change

The microcosm N-system includes material circulation, energy flow, and microbial interactions, which are the foundations of natural ecosystems. It is possible to set the culture conditions of the microcosm to reflect the influence of global climate change, using the microcosm to diagnose the influence of climate change on Earth's ecosystems. In this section, the effects of climate change on ecosystems are investigated using the microcosm N-system.

7.15.1 Global Warming

Global warming is one of the most important environmental problems because of the seriousness of its effects on ecosystem structure, including upon energy flow, material circulation, and biological interactions. The Intergovernmental Panel on Climate Change (IPCC) reported in 2007 that 40% of all wildlife species on Earth will become extinct with a 4 °C increase in global atmospheric temperatures. However, information about the mechanisms underlying the relationship between global warming and the succession of biological species in aquatic environments has been insufficient to establish countermeasures, either academically or administratively. For this reason, experimental data are of extreme importance and must be obtained rapidly and in great abundance. We investigated the effects of temperature on microbial community function and structure using the experimental flask-sized microcosm.

A flask-sized microcosm was cultured at 25 °C under 2400 lux (12 h. L/12 h. D) and without stirring for use as the control system. To assess the influence of temperature, 30 °C, 35 °C, and 40 °C were used as the high-temperature conditions of global warming, and 10 °C and 20 °C were used as the low temperature conditions of global cooling. The vessel for cultivation was a 300 mL Erlenmeyer flask, and 200 mL of Taub's basal medium was poured; the concentration of the polypeptone as a substrate was adjusted to 100 mg/L. Plankton observation using an optical microscope was conducted from the start of culturing, on days 0, 2, 4, 7, 14, 16, 18, 20, 23, and 30. The kind of change observed in a flask under each temperature condition was used as the structural parameter. Additionally, the DO concentration was measured consecutively from the 16th day of the culture period onward to evaluate the functioning of the ecosystems by using the P/R ratio as the functional parameter.

In high-temperature conditions (35 °C and 40 °C), zooplankton consumers disappeared, whereas all microbial species maintained coexistence at 30 °C. The no-effect temperature in the microcosm was evaluated as less than 35 °C from the viewpoint of ecosystem structure. With increases in temperature, the activity of the system increased, and the P/R ratio was stable. However, the DO amplitude damped with the extinction of the microcosm at 35 °C. The maximum environmental temperature to have no effect on the microcosm, judging from the functional parameter, was evaluated to be less than 35 °C. At the lowest temperature (10 °C), the consumer population of the microcosm decreased, the activity of the ecosystem decreased, and the rate of succession slowed. It was shown that the influence of high temperatures was particularly strong with respect to the consumers, and the entire system eventually collapsed (Figs. 7.43 and 7.44).

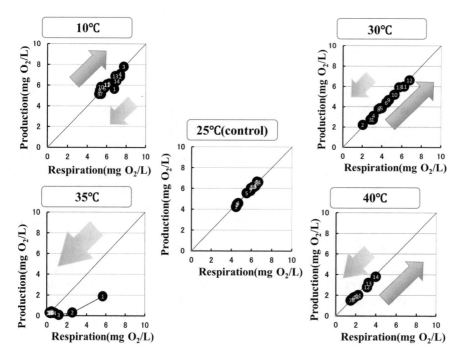

Fig. 7.43 Time course of P/R ratio in each culture temperature

→: activity of microcosm system increased in time course

→: activity of microcosm system decreased in time course

*The activity in each temperature was increased initially and then decreased at 10 °C with zooplankton decrease, decreased initially and then increased at 30 °C with no damage of microbiota, decreased at 35 °C with zooplankton disappearance, and decreased initially and then increased at 40 °C with zooplankton disappearance

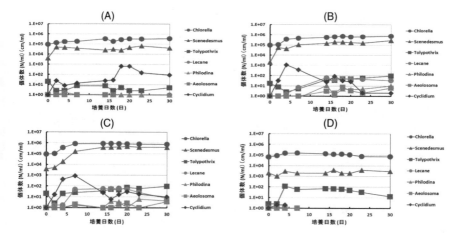

Fig. 7.44 Time course of population in each culture temperature. (**a**) 10 °C (low temp.). (**b**) 25 °C (normal temp.). (**c**) 30 °C (high temp.). (**d**) 40 °C (high temp)

7.15.2   Acid Rain

To assess the effect of acid rain on ecosystems, an HCl addition investigation was conducted by Dr. Sugiura. The microcosm was cultured with the addition of HCl at concentrations of 10 µmol, 20 µmol, 35 µmol, 75 µmol, 100 µmol, and 200 µmol at the start of cultivation (day 0) to investigate the effects of acid rain on ecosystems. The pH values were 5.40, 4.92, 4.45, 3.78, 3.60, and 3.15. Except for the 10–35 µmol of HCl addition systems, the time series of the P/R ratio exhibited different behaviors between the addition and control systems. Some species of microorganisms in the microcosm went extinct with 100 µmol of added HCl. The difference in the time series of the P/R ratio from the control system increased according to the increase in the amount of HCl added until the 75 µmol concentration was reached. Changes in the P/R ratio over time without the extinction of microorganisms was considered to be caused by changes in the biological interactions and balance with the addition of HCl. The fact that microorganisms in the microcosm N-system could resist changes in pH is considered to result from its biodiversity and suggests that it is useful as a model of natural ecosystems (Sugiura 1993).

Literature Cited

Beyers RJ. The metabolism of twelve aquatic laboratory microecosystem. Ecol Monogr. 1963;33:281–306.

Kurihara Y. Studies of succession in a microcosm, science report of Tohoku University, 4th Ser. (Biology), No.37, 1978a. p. 151–60.

Kurihara Y. Studies of the interaction in a microcosm, science report of Tohoku University, 4th Ser. (Biology), No.37, 1978b. p. 161–77.

Chapter 8
Coefficient of Assessment for the Microcosm System

Kunihiko Kakazu, Kakeru Ruike, and Kazuhito Murakami

Abstract In this chapter, the concept of the coefficient of assessment and the limitations of the correlative analysis of the microcosm and mesocosm are described.

8.1 Coefficient of Assessment

A conceptual diagram of the unevenness of the NOEC in the same test material in the microcosm test is shown in Fig. 8.1. When ecosystemic sensitivity is assumed to be normally distributed, and because the sensitivity varies according to the exposure of the ecosystem to the same chemical substance and demand, the standard deviation (σ) of the sensitivity distribution from experimental data is equal to 0.51. In other words, the density that only σ moved the mean is equivalent to density to protect ecosystemic 84%. When the standard deviation of the regression lines of both the correlation of the NOEC in the microcosm test and the NOAEC in the outdoor ecosystem test (i.e., the mesocosm test) is equal to 0.57, the range over which only σ moved the regression line downward may account for 84% of the chemical substance. Meanwhile, a conceptual diagram of the unevenness of the NOEC in a different test material is shown in Fig. 8.2.

The validity of the toxicity evaluation using the microcosm test was examined from the relationship between the NOEC in the microcosm and the average value of the NOEC in the natural ecosystem. It was indicated that 23 out of 26 substances fell within the confidence interval of the regression line, and a strong correlation between

K. Kakazu
Foundation for Advancement of International Science, Tsukuba, Ibaraki, Japan
e-mail: kakazu@fais.or.jp

K. Ruike
University of Tsukuba/Foundation for Advencement of International Science, Tsukuba, Ibaraki, Japan
e-mail: ruike@fais.or.jp

K. Murakami (✉)
Chiba Institute of Technology, Narashino, Chiba, Japan
e-mail: kazuhito.murakami@p.chibakoudai.jp

© Springer Nature Singapore Pte Ltd. 2020
Y. Inamori (ed.), *Microcosm Manual for Environmental Impact Risk Assessment*,
https://doi.org/10.1007/978-981-13-6798-4_8

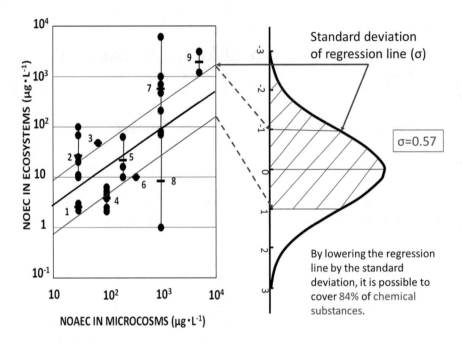

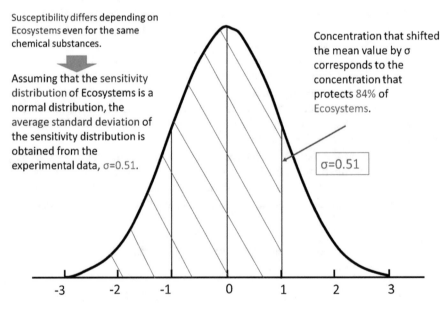

Fig. 8.1 Variation of NOEC on same chemical agent

*When ecosystemic sensitivity is assumed to be normally distributed, and because the sensitivity varies according to the exposure of the ecosystem to the same chemical substance and demand, the standard deviation (σ) of the sensitivity distribution from experimental data is equal to 0.51, that is, the density that only σ moved the mean is equivalent to density to protect ecosystemic 84%. When the standard deviation of the regression lines of both the correlation of the NOEC in the microcosm the microcosm and experimental ecosystem (i.e., natural ecosystem) was observed. The lower limit of the confidence interval is a straight line obtained by dividing the average value of the NOEC of the natural ecosystem by the uncertainty factor (coefficient of assessment = 200), obtained by considering the difference in ecosystem sensitivity. Only the lower limit is needed to predict the NOEC in the natural ecosystem. If the correlation coefficient is greater than this lower limit (i.e., plots above the lower limit line), the concentration obtained from dividing the NOEC measured using the microcosm by the coefficient of assessment (i.e., the predicted NOEC) will be lower than the predicted NOEC (PNOEC) of the natural ecosystem (i.e., the mesocosm experiment); this means that it can predict natural ecosystems. In this study, most of the measured substances were contained in this numerical region (i.e., above the lower limit line). However, although substances exceeding the upper limit can predict the PNOEC, the value is smaller than the NOEC of the natural ecosystem. Only one substance out of 26 substances exceeding the upper limit could be accurately predicted using the microcosm test, regardless of the substance. Additionally, a comparison of the PNOEC of the natural ecosystem and the microcosm NOEC, as well as various techniques, are shown in Fig. 8.3. The microcosm test has equivalent PNOEC and NOEC, there is little variation, and precision is high. Moreover, an impact statement is possible for ecosystemic sensitivity at approximately the same level, in comparison with certain ECETOC and HC5 methods that are more conventionally used.

8.2 Comparison and the Interface of the Microcosm and Natural Ecosystem (Mesocosm)

When the validity of the microcosm test was considered, it was shown that 23 materials fell within the confidence interval of the regression line out of 26 total materials, and the strength of the correlation between the experimental ecosystem (microcosm) and the natural ecosystem (mesocosm) was illustrated by the correlations between the means of the NOEC of the microcosm N-system and the NOEC of the natural ecosystem (Fig. 8.4). Most materials that were measured in this study fit into this domain (i.e., above the lower limit), as described in Sect. 8.1. However, for the material value that was greater than the upper limit, the value was less than that of the natural ecosystem NOEC, although it could be used to predict the PNOEC. The characteristics of the chemical substance used in the microcosm test are shown in Table 8.1.

Two substances that fall outside of the confidence interval of the correlation between the microcosm NOEC and the average NOEC of the natural ecosystem

←───

Fig. 8.1 (continued) test and the NOAEC in the outdoor ecosystem test (i.e., the mesocosm test) is equal to 0.57, the range over which only σ moved the regression line downward may account for 84% of the chemical substance

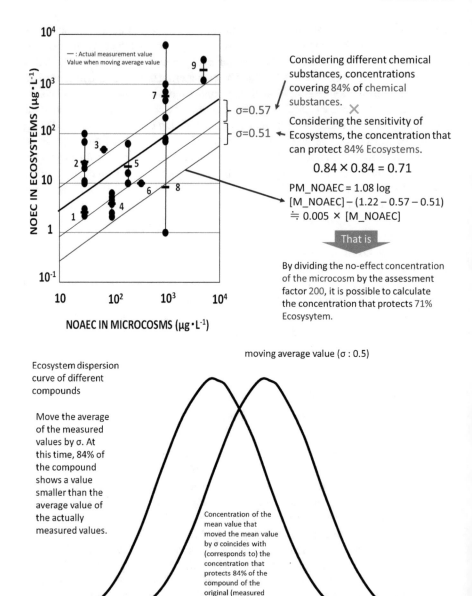

Considering different chemical substances, concentrations covering 84% of chemical substances. ×

Considering the sensitivity of Ecosystems, the concentration that can protect 84% Ecosystems.

$$0.84 \times 0.84 = 0.71$$

PM_NOAEC = 1.08 log [M_NOAEC] − (1.22 − 0.57 − 0.51) ≑ 0.005 × [M_NOAEC]

That is

By dividing the no-effect concentration of the microcosm by the assessment factor 200, it is possible to calculate the concentration that protects 71% Ecosysytem.

Ecosystem dispersion curve of different compounds

Move the average of the measured values by σ. At this time, 84% of the compound shows a value smaller than the average value of the actually measured values.

moving average value (σ : 0.5)

Concentration of the mean value that moved the mean value by σ coincides with (corresponds to) the concentration that protects 84% of the compound of the original (measured value) dispersion curve.

Fig. 8.2 Variation of NOEC on different chemical agent

(mesocosm) are characterized by high soil adsorption. Possible explanations of this pattern include paraquat (an herbicide), which exhibits a weak correlation and is overestimated; chlorpyrifos (an insecticide), which also exhibits a weak correlation

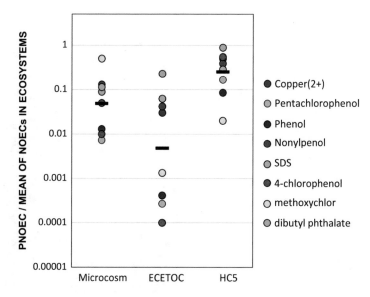

Fig. 8.3 Comparison of PNOEC obtained by various assessment test method
—; mean ●; PNOEC/mean of NOECs in ecosystems
*The microcosm test has equivalent PNOEC and NOEC, there is little variation, and precision is high. Moreover, an impact statement is possible for ecosystemic sensitivity at approximately the same level, in comparison with certain ECETOC and HC5 methods that are more conventionally used

and is underestimated; or experimental conditions in the field test (mesocosm), such as soil adsorption and continuous vs. intermittent administration. In the future, further investigation into the causality of the two outliers will be necessary. For example, to investigate chlorpyrifos in the natural experiment (mesocosm test) with an intermittent administration system, the following two approaches could be compared:

(i) Single administration with soil: due to adsorption in the soil and hydrolysis, the concentration in water sharply decreases, but a high soil/water interface concentration is continuously maintained by desorption from the soil.

(ii) (ii) Single administration without soil: there is a first-order decrease in the concentration of water in the microcosm, and, in turn, the toxicity is low.

Paraquat is known to be easily adsorbed in the soil. The value of paraquat multiplied by the coefficient of assessment can predict the conditions in the natural ecosystem, but there is no correlation between the microcosm test and the mesocosm test. Based on these results, even though the environmental risk assessment of chemical substances can be evaluated in the microcosm, it is necessary to establish an OECD international standard by defining the limitations of the test. The results obtained in this study were presented at the 3rd JACI LRI research meeting on August 28, 2014, and are shown in this book.

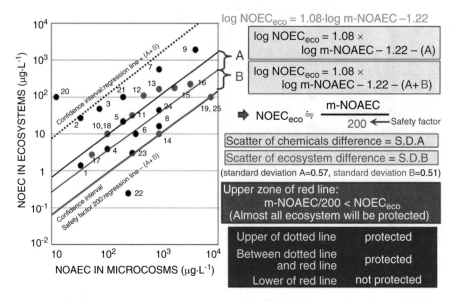

Fig. 8.4 Relationship between m-NOEC and NOEC in ecosystem
●, mean of NOECs; —, regression line; …, upper limit of confidence range; —, lower limit of confidence range; 1, Cd^{2+}; 2, atrazine; 3, PCP; 4, Cu^{2+}; 5, Zn^{2+}; 6, ammonia; 7, LAS; 8, 3,4-DCA; 9, phenol; 10, nonylphenol; 11, Mancozeb; 12, 2,3,4,6-tetrachlorophenol; 13, TMAC; 14, alachlor; 15, AE; 16, SDS; 17, DCMU; 18, linuron; 19, carbendazim; 20, paraquat; 21, 4-chlorophenol; 22, chlorpyrifos; 23, methoxychlor; 24, dibutyl phthalate; 25, 2,4-dichlorophenol; standard deviationA, variation of chemicals; standard deviationB, variation of natural ecosystem

Based on our results, we classified a wide range of chemical substances based on their respective characteristics and continued research and verification, accumulating data that can be scientifically supported. To establish a practical ecological impact assessment method for the OECD international standardization of the microcosm test, the following specific issues were addressed: (1) confirmation of the reproducibility of the test methods using a blind test performed by three independent organizations targeting the same substance, (2) verification of versatility using a ring test (performed by five organizations) in domestic organizations, (3) building a practical foundation for the microcosm NOEC database, (4) building the foundation for a public test method through the completion of this test manual, and (5) verification of the utility of the test method by constructing an environmental NOEC prediction method. These steps were deemed necessary, and the findings of previous studies have been clarified through the application of this approach.

For steps 1–4, based on the characteristics of the chemical substances, a comparative analysis of reproducibility when a substance was added at the beginning of culturing and during the steady state was performed as necessary. The aims of generalization are to examine and strengthen the efficiency of the experiment with respect to the timing of substance addition, the concentration range, and the precision of the appropriate settings. The aim of step 5 was to expand the database of the

Table 8.1 Chemical characteristics of test agent in microcosm test

Chemicals	ECOSAR class	Log Kow	Most sens. species
Cadmium (2+)	Inorganic	–	Daphnid
Copper(2+)	Inorganic	–	Daphnid
Zine(2+)	Inorganic	–	Fish
Ammonia	Inorganic	0.23	Daphnid
Pentachlorophenol	Phenols	4.74	Fish
Nonylphenol	Phenols	5.99	Fish
4-chlorophenol	Phenols	2.16	Daphnid
2,3,4,6-tetrachlorophenol	Phenols	4.09	Daphnid
Phenol	Phenols	1.51	Daphnid
2, 4-dichlorophenol	Phenols	2.80	Daphnid
LAS	Surfactants–anionic	3.80	Daphnid
SDS	Surfactants–anionic	2.42	Algae
TMAC	Surfactants–cationic	1.22	Daphnid
AE	Surfactants–nonionic	2.85	Daphnid
DCMU	Substituted ureas	2.67	Algae
Linuron	Substituted ureas	2.91	Algae
Lindane	Neutral organics	4.26	Daphnid
Methoxychlor	Neutral organics	5.67	Fish
Atrazine	Triazines, aromatic	2.82	Algae
Chlorpyrifos	Esters, monothiophosphates	5.11	Daphnid
Mancozeb	Thiocarbamate, Di	0.62	Daphnid
3,4-dichloroaniline	Anilines	2.37	Daphnid
Alachlor	Haloacetamides	3.37	Algae
Dibutyl phthalate	Esters	4.61	Fish
Carbendazim	Imidazoles, carbamate esters	1.55	Daphnid
Paraquat	Not related	2.71	Algae

NOECeco of the mesocosm test and the m-NOEC of the microcosm test. The applicable range of the microcosm, a method for determining the coefficient of assessment, taking into consideration the dispersion of each chemical substance from the correlation matrix, and the difference in susceptibility for each ecosystem were established. Further, the limitations of the microcosm test proposed in this study are based on the assessment of chemical properties. At the same time, the conceptual underpinning of extrapolating to natural ecosystems to establish a verified technique and database, which is the basis of the analysis in this book, were further expanded upon.

Expansion of the applicable range of the microcosm test methods and optimization of such methods are examined in this manual, and an OECD international standardization test method of this microcosm test method was established by expanding the database necessary for determining the coefficient of assessment. In the future, further examination of the properties of the chemical substances used, as well as grouping and promoting additional research and verification, will be needed so that the microcosm test is established as a practical ecosystem impact assessment method, replacing the mesocosm test method. As microcosms are expected to be highly correlated with natural ecosystems, it is assumed that more realistic estimates of the NOEC may be obtained when compared with the current method (i.e., single-species evaluation). Currently, it remains difficult to omit test methods using algae, crustaceans, and fish in ecological impact assessments of chemical substances, as typified by the WET test. However, by using an ecosystem model with a food chain and energy flow in parallel, it is possible to accumulate knowledge concerning the decomposability and persistence of chemical substances, restoration of ecological functions, and collapse. In other words, it is expected that the advantages of the microcosm will be evaluated by establishing a method of quantifying and evaluating ecosystem impacts. The WET test is a method by which toxicity, including multiple effects, may be evaluated. It does not evaluate the toxicity of individual chemical substances, but rather it targets waters that contain multiple chemical substances. However, when species from different niches in the food chain are used in the assay, it is a single-species test. It must be noted that a problem remains in that material circulation, energy flow, and biological interactions, which are the basic elements of an ecosystem, are not incorporated in the test method under these conditions. The microcosm, which is a system of multiple coexisting species, evaluates the risk of chemical substances at the ecosystem level. The safety factor obtained by this method also differed from the conventional method, according to the correlation between the mesocosm test and the microcosm test. With the accumulation of further data, it will be possible to calculate realistic values of the PNOEC for natural ecosystems.

Literature Cited

Suzuki M, Utsumi H. Bioassey –risk management of water environment. Tokyo: Kodansha Scientific; 1998. 270pp.

The Japanese Society of Environmental Toxicology. Handbook of ecosystem assessment test –environmental risk assessment of chemicals, Tokyo. Asakura SHoten; 2003. 349pp.

Urano K, Matsuda H. Principles and methods for ecosystem risk management. Tokyo: Ohmsha; 2007. 209pp.

Utsumi H, Nakamuro K. Bioassey and bio-informatics, fpr environmental assessment and medicinal sciences. Tokyo: Kougaku-Tosho Publishers Ltd; 2011. 214pp.

Wakabayashi A. Chemicals and ecotoxicity, Tokyo. Japan Environmental Management Association for Industry; 2004. 486pp.

Yoshida K, Nakanishi J. Basic environmental risk analysis. Tokyo: Tokyotosho; 2006. 234pp.

Chapter 9
Application to the Whole Effluent Toxicity Test

Yuhei Inamori, Ryuhei Inamori, Kazuhito Murakami, and Yuzuru Kimochi

Abstract In Japan, evaluation methods have been developed for the introduction of the WET test. However, the WET test is a test for a single species, as typified by *Daphnia magna* and *Danio rerio*. For this reason, biological interactions are not considered, and it is possible that this will cause impact assessments to differ from the conditions in natural ecosystems. Therefore, we aimed to further develop and verify the WET test using the microcosm. In addition to the single-species test using species belonging to different niches (ecological roles), such as fish, crustaceans, and algae, the microcosm test was conducted as a multiple-species test correlated with a natural ecosystem. It was necessary to construct a comprehensive test method based upon the concepts of ecological impact assessments. The microcosm test can also be incorporated as an expanded test along with a single-species test if necessary.

9.1 Necessity of the Microcosm Test for WET

Impact assessments at the ecosystem level utilizing aquatic model ecosystems, including a microcosm, are well understood, but focus on the WET test method utilizing a microcosm as a wastewater management method is needed. In parallel with a single-species test using species belonging to different niches, such as fish, crustaceans, and algae, microcosm tests are positioned as multiple-species tests. They may be correlated with natural ecosystems, and it is essential to construct these systems as comprehensive test methods based on ecological impact assessment.

Y. Inamori (✉) · R. Inamori
Foundation for Advancement of International Science, Tsukuba, Ibaraki, Japan
e-mail: y_inamori@fais.or.jp; r_inamori@fais.or.jp

K. Murakami
Chiba Institute of Technology, Narashino, Chiba, Japan
e-mail: kazuhito.murakami@p.chibakoudai.jp

Y. Kimochi
Center for Environmental Science in Saitama, Kazo, Saitama, Japan
e-mail: kimochi.yuzuru@pref.saitama.lg.jp

© Springer Nature Singapore Pte Ltd. 2020
Y. Inamori (ed.), *Microcosm Manual for Environmental Impact Risk Assessment*,
https://doi.org/10.1007/978-981-13-6798-4_9

The WET test is a method that evaluates toxicity, including multiple effects, because it is targeted at test waters containing multiple chemical substances rather than being used for evaluations of the toxicity of individual chemical substances. However, although we use individual species (fish, crustaceans, and algae) located in different niches in the food chain for the assay, it is a single-species test only. It is not a test method that accounts for material cycles, energy flow, or biological interactions, which are the basic elements of natural ecosystems. The microcosm is a system that includes multiple species in coexistence and contains material circulation, energy flow, and biological interactions. Therefore, the microcosm makes it possible to evaluate the risks associated with added chemical substances at the ecosystem level. The safety factors for the ecological effects of chemical substances obtained by the microcosm test were supported by the high correlation observed between the microcosm test and mesocosm (natural ecosystem) test obtained from the Environmental Research Comprehensive Promotion Fund (2009–2011) and the New LRI (2012–2014). Therefore, we determined that it is possible to use the microcosm test to evaluate the effects of chemical substances in an ecosystem with higher precision and with less dispersion than in the conventional single-species tests, such as the HC5 method. Consequently, it is possible to calculate more realistic values of the PNOEC for natural ecosystems using the microcosm.

Microcosms are highly correlated with natural ecosystems and can be regarded as models containing the basic components and characterized by the basic principles of natural ecosystems. From this fact, it is considered that a more realistic estimate of the NOEC can be obtained using a microcosm as compared with the current single-species test that does not account for material cycling, energy flow, or biological interactions. In the ecosystem impact assessment of wastewater containing various chemical substances, comprehensive tests were conducted to introduce the ecosystem model, a multiple-species test with a food chain and energy flow, simultaneously with the single-species test. A series of test methods were constructed, making it possible to evaluate the degradability and persistence of chemical substances contained in wastewater and the collapse and restoration of ecosystem functions.

The microcosm test can be performed at a lower cost than the WET test with commonly used single species (fishes, crustaceans, and algae). The microcosm has already been subcultured for nearly 40 years and is easy to use by researchers accustomed to environmental microorganism testing who can master the microbial culturing procedures. It is also easy to use after short-term training. Additionally, a series of tests are possible without using expensive culturing tools or measurement instruments.

The position of the microcosm test within the WET test is shown in Fig. 9.1. The foundation of a certain ECETOC method and the HC5 method conventionally includes the evaluation of independent chemical substances in toxicity tests using single species, and the current WET test also uses single species for toxicity tests on the composition of chemical substances. The interactions between species constitute the ecosystem and drive the cycling of materials and energy flow; no single species is

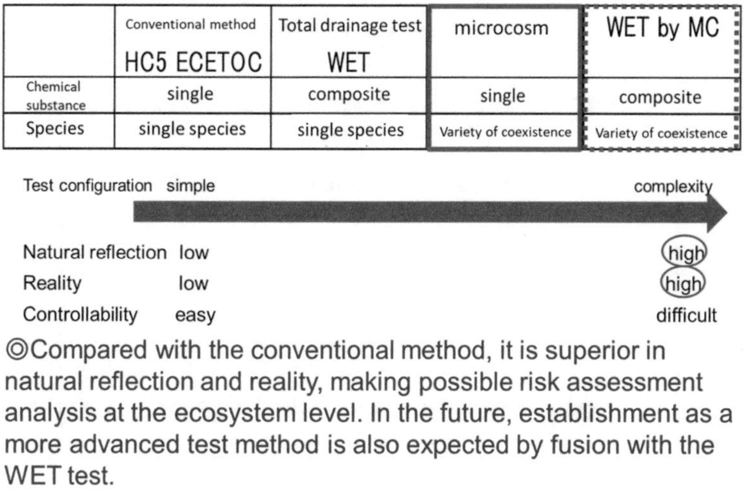

	Conventional method HC5 ECETOC	Total drainage test WET	microcosm	WET by MC
Chemical substance	single	composite	single	composite
Species	single species	single species	Variety of coexistence	Variety of coexistence

Test configuration simple   complexity

Natural reflection low   (high)
Reality   low   (high)
Controllability   easy   difficult

©Compared with the conventional method, it is superior in natural reflection and reality, making possible risk assessment analysis at the ecosystem level. In the future, establishment as a more advanced test method is also expected by fusion with the WET test.

Fig. 9.1 Placement of microcosm test for WET

able to exist alone. It is therefore insufficient to evaluate the influence of added chemical substances at the ecosystem level using a single-species test. However, the model ecosystem consists of combinations of species, and multiple microcosm tests contain the basic elements of ecosystems, such as biological interactions, cycling of materials, and energy flow between species. Using the microcosm, it is possible for impact assessments at the ecosystem level to reflect the natural ecosystem, advancing the WET test as well evaluating single chemical substances.

9.2 Preparation and Addition of Test Effluent

9.2.1 Preparation of Test Effluent

According to the testing procedure described in the 1997 Sewage Examination Method (Japan Sewage Works Association, Volume III Biological Examination, Chap. 1 Biological Examination, Sect. 10, Ecosystem Impact Assessment Testing), the ratio of TP medium and test effluent should be set to 0%, 25%, 50%, 75%, or 100%. Remove contaminants by filtering with a 0.45 μm size mesh filter, and add the filtrate to the microcosm. The time of addition is at the beginning of the growth phase (i.e., day 0 of cultivation) or during the stationary phase (16th day of cultivation). The water quality of the test effluent should be analyzed if necessary. The procedures of microcosm-WET test are shown in Figs. 9.2 and 9.3.

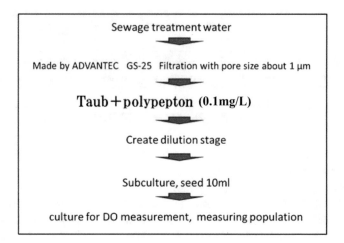

Fig. 9.2 Procedure of microcosm-WET test

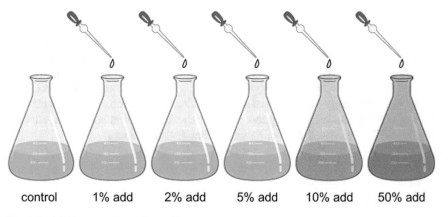

control   1% add   2% add   5% add   10% add   50% add

Fig. 9.3 Addition procedure of test effluent

9.2.2 Addition of Test Effluent

Addition of test effluent to the microcosm is conducted on day 0, when cultivation of the microcosm begins, or on the 16th day, when the microcosm has reached a stationary phase. For experiments with day 0 additions of chemical substances, the effect of the test effluent on the microcosm and the constituent microbiota may or may not allow a stationary phase to be reached (i.e., a stable ecosystem may or may not be established). For experiments in which chemical substances were added on the 16th day, the effect of the test effluent on the microcosm and the constituent microbiota may or may not allow for the retention of species diversity (i.e., a stable ecosystem may or may not be recovered). For both addition methods, the structural parameter (microbial population) and the functional parameter (P/R ratio) should be measured to evaluate the effects of the test effluent.

In the conventional microcosm experiments, chemical addition is usually performed on the 16th day after the cultivation began, when the system stabilizes. In the WET test using the microcosm, the medium in the flask must be exchanged for wastewater, which is then added. At that time, it is necessary to separate the organisms and the culture medium, and a fresh medium containing wastewater is added instead of the extracted medium. The addition of the attached microcosm standard medium increases the number of individuals in the steady state. This is because organic substances that feed microorganisms are already consumed. In this case, it is impossible to compare it with the nonadditive system (control), so an appropriate medium was developed with reference to the amount of organic matter added at the time of wastewater addition.

Addition of a substrate was conducted by replacing the culture medium in the flask with wastewater. A fresh medium containing wastewater was added as a substitute for the extracted medium. However, when the standard microcosm medium was added as a new medium, the number of individuals in the steady state increased. Therefore, as a new medium, the amount of peptone in the medium was adjusted to 0 mg/L, 5 mg/L, 10 mg/L, 20 mg/L, and 40 mg/L and then added into the medium. The impact of each culture medium was assessed from both the P/R ratio as the functional parameter and the abundance of microbiota as the structural parameter in the microcosm (Murakami et al. 2017). The environmental impact and ecological risk were estimated by comparing the treated microcosms with the no-addition system (control) in both assessment methods.

With respect to the structural parameters, the abundance of *Cyclidium glaucoma*, a ciliate primary predator, was greatly increased by increasing the amount of peptone in the medium; there was no major change in the abundance of other species in the microcosm. From the P/R ratio, which is the functional parameter, no change was observed when compared to the no-addition system (control) as shown in Fig. 9.4. From this, the system was considered stable unless wastewater was added. Statistical analysis of the DO concentration was performed. The slope (a) and the coefficient of variation (cv) were obtained assuming that the microcosm was normally distributed. If they fell within the range of $\pm 34.13\%$ for each, it was considered that there was no influence. Since the range of $\pm 34.13\%$ of the slope of the DO was $0.0028 \leqq a \leqq 0.0056$, it was considered as having no influence on the medium containing only 5 mg/L. Additionally, since the range of $\pm 34.13\%$ of the coefficient of variation of the DO was $0.10 \leqq cv \leqq 0.20$, it was considered as having no effect on the medium containing 0 mg/L to 20 mg/L. Therefore, a medium containing 5 mg/L of peptone was considered appropriate for adding wastewater in the microcosm-WET test (Fig. 9.4).

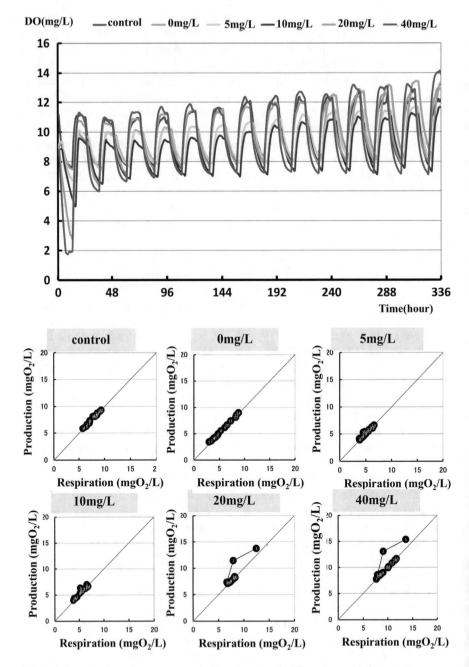

Fig. 9.4 Investigation of proper concentration of peptone in TP medium added to microcosm

9.3 Example of the Microcosm-WET Test

Examples of WET tests using a microcosm are presented below for sewage-treated domestic and plant drainage water, rice-washed effluent, paddy field effluent, golf course effluent, factory effluent, and landfill leachate.

9.3.1 Sewage-Treated Domestic Drainage

Sewage-treated water from domestic drainage that was suction filtered using a 1 μm diameter membrane (ADVANTEC, GS-25) was added at the initial stage of microcosm cultivation (day 0), and the addition concentrations were set to 0% and 80% in the microcosm medium. By comparing the 80% addition of the sewage-treated domestic drainage with the control system (0% addition), the ecological effects were assessed relative to the total toxicity. As a result, in the sewage-treated domestic drainage, the extinction of *Aeolosoma hemprichi* was observed, whereas the decreases observed in *Lecane* sp., *Philodina erythrophthalma*, *Tolypothrix* sp., and *Scenedesmus quadricauda* tended to recover. Even in the evaluation using P and R, the initial P and R values tended to decrease in the sewage-treated domestic drainage more than in the control system, and it can be inferred that respiration and photosynthesis were inhibited. With respect to the classification patterns of ecosystem influence (i.e., maintenance, recovery, metastasis, and disintegration of the system), the sewage-treated domestic drainage (80%) was classified as a metastasized system.

9.3.2 Sewage-Treated Plant Drainage

Sewage-treated water from plant drainage that was suction filtered using a 1 μm diameter membrane (ADVANTEC, GS-25) was added at the initial stage of microcosm cultivation (day 0), and the addition concentrations were set to 0%, 20%, 40%, and 80% in the microcosm medium. The ecological effects were evaluated from the total toxicity by adding a dilution stage (0%, 20%, 40%, and 80%) of sewage-treated plant drainage, which was highly toxic. As a result, at 80%, all micro-animals and *Tolypothrix* sp. went extinct, and, at 40%, only *Cyclidium glaucoma* maintained its abundance, while the abundance of *Tolypothrix* sp. decreased. At 20%, *Aeolosoma hemprichi* became extinct, and the populations of other species, except for *Chlorella* sp., decreased. Even in the evaluation using P and R, as shown in Fig. 9.5, P and R transitions were similar to those in the control system, and the P/R ratio was observed to gradually decline. From the results obtained when the sewage-treated plant drainage was diluted in a stepwise manner, when the influence patterns were

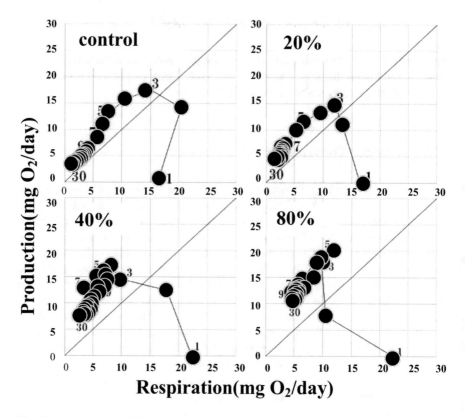

Fig. 9.5 Time course of P/R ratio in sewage-treated plant drainage added microcosm

classified, it was classified as being in a state of system collapse with 80% sewage-treated plant drainage and a system transition with 40% and 20% sewage-treated plant drainage. Additionally, when diluted to 20%, it was nearly identical to the control system. The sewage derived from plant drainage was demonstrated to have a substantial effect on the microcosm. The influence of sewage-treated water from domestic or plant drainage on the microcosm was also clearly confirmed from the appearance of the microcosm during the culturing process. For sewage-treated water, the substances that may be influenced by precise values of heavy metals and other chemicals were analyzed, and it was suggested that copper, nickel, and chlorothalonil may influence the added substances (Fig. 9.5).

9.3.3 Rice-Washed Effluent

A large amount of nitrogen and phosphorus are included in cleansed rice drainage. These are causative agents of eutrophy and lead to red tides and blue tides—

outbreaks of algae—if discharged carelessly. In this section, the WET test was used to investigate the influence of rice-washed effluent on the microcosm (i.e., the model of the natural ecosystem). Additionally, the influence of rice-washed effluent on the ecosystem in which it was discharged was determined based on the toxicity units (TUs) of the m-NOEC.

Rice-washed wastewater that was suction filtered using a 0.45 μm diameter membrane (ADVANTEC, GS-45) was added at a stable stage 16 days after microcosm cultivation began, and the addition concentrations were set to 0%, 5%, 10%, 20%, 40%, and 80% of the microcosm medium. The impact of rice-washed effluent was estimated from both the P/R ratio (Sugiura 2010) as the functional parameter and the abundance of microbiota (Murakami et al. 2016) as the structural parameter in the microcosm. The environmental impact and ecological risk were estimated by comparing the microcosm with the no-addition system (i.e., the control system) using both assessment methods.

As the addition concentration increased in comparison with the control system, the amplitude of the DO increased, and the rice-washed effluent raised the bioactivity in the microcosm. The time series of the P/R ratio was stable, and both phytoplankton (producers) and zooplankton (consumers) were assumed to multiply together with all addition concentrations (i.e., 0%, 5%, 10%, 20%, 40%, and 80%) in the microcosm. Because $\pm 34.13\%$ of the control system values of the slope a in the regression equation were $0.0025 \leq a \leq 0.0051$, it was determined that there was no influence in the 10% addition system. Because $\pm 34.13\%$ of the control system values of the coefficient of variation cv exhibited amplitudes wherein $0.099 \leq cv \leq 0.20$, it was determined that all addition concentrations of rice-washed effluent had an impact on the ecosystem. Based on the functional parameter (P/R ratio), all addition concentrations also had an influence on the ecosystem. However, there was no influence noted for any addition concentrations based upon the structural parameter because no population declines or extinctions were observed.

A microcosm-WET test for rice-washed effluent was conducted, and the influence of the rice-washed effluent on the m-NOEC was less than 5% in both the functional and the structural parameters. Furthermore, the toxicity unit (TU) was determined to be 20, so dilution by more than 20 times was necessary. It is also necessary to examine the consistency using statistical analyses, and a branching-type analysis of variance was used to validate these findings; this simple statistical technique is described in the microcosm test method manual.

9.3.4 Paddy Field Effluent

Paddy field effluent, after pesticide spraying, was suction filtered using a 0.45 μm diameter membrane (ADVANTEC, GS-45) and added at a stable stage on the 16th day after microcosm cultivation began. The addition concentrations were set to 0%, 25%, 50%, and 100% of the microcosm medium. The abundance of *Cyclidium glaucoma* decreased in the 50% addition system, as shown in Figs. 9.6 and 9.7, and became extinct in the 100% addition system. Therefore, it was determined that a

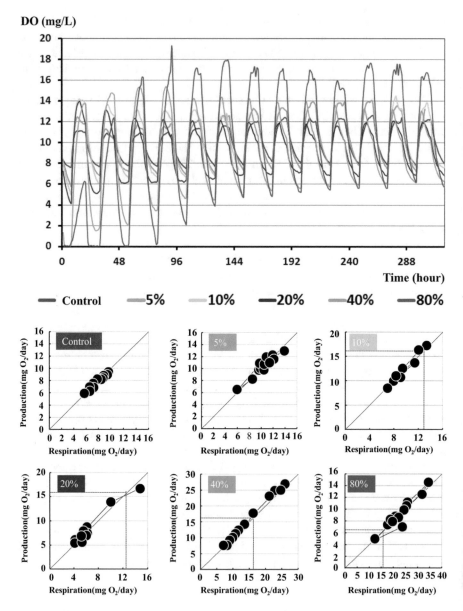

Fig. 9.6 Time course of DO and P/R ratio in rice-washed effluent added microcosm-WET test

dilution of more than double the concentration was necessary to maintain the system. The m-NOEC of the structural parameter was estimated as 1–10%; more specifically, it was found that there would be some influence on the ecosystem at any addition concentration without being diluted more than 10 times (Fig. 9.7).

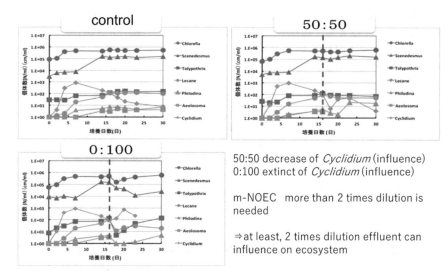

50:50 decrease of *Cyclidium* (influence)
0:100 extinct of *Cyclidium* (influence)

m-NOEC more than 2 times dilution is needed

⇒ at least, 2 times dilution effluent can influence on ecosystem

Fig. 9.7 Time course of population in paddy field effluent added microcosm-WET test

9.3.5 River Water Containing Germicides and Pesticides

River water that flowed after dispersion and was sampled for a day was targeted for evaluation before spraying the insecticide, DEP 50%, and the sterilizer, tricyclazole 20%, over the rice field. An inorganic salt solution and 5 mg/L of polypeptone and Taub's basal medium was added to the river water, which was then filtered and sterilized. An inorganic salt solution and 50 mg/L of polypeptone and Taub's basal medium were poured by 20 mL amounts into the experimental system. Finally, a TP_{50} nutrient medium was used for the control system, and the microcosm was subcultured in a TP_{50} nutrient medium with a microcosm seed at 2800 lux (L/D = 12 h/12 h), 25 °C, without stirring.

In the river water before pesticide dispersion, an increase was seen in the bacteria as in the control system until microcosm culturing began on the fourth day, but it tended to decrease sharply after the seventh day. Additionally, with respect to the producer, a notable increase was seen in the abundance of *Chlorella* sp. However, in *Tolypothrix* sp., an increase in abundance was controlled, and, for the predators *Cyclidium glaucoma*, *Lecane* sp., *Philodina erythrophthalma*, and *Aeolosoma hemprichi*, survival was possible, but the increase in *Aeolosoma hemprichi* was half that in the control system. Therefore, there were many things that dispersed, and a patchy cluster was characteristic of this microcosm in which high predation activity of *Philodina erythrophthalma* on *Chlorella* sp. was observed. The possibility that even the river water before the pesticide dispersion had an influence on the microcosm ecosystem was thus suggested.

When microcosm culturing with the river water began soon after pesticide dispersion, an increase was seen in the bacteria, as in the control system. Additionally, an increase was also seen in the abundance of the producer, *Chlorella* sp., as was observed before pesticide dispersion, but the increase in *Tolypothrix* sp. was controlled, and the survival of *Lecane* sp. and *Philodina erythrophthalma* was confirmed. *Cyclidium glaucoma*, which predominated early in the culture of the control system, and *Aeolosoma hemprichi* were not observed again. As for the patchy structure, the formation of clusters decreased after pesticide dispersion relative to before pesticide dispersion, and abundant, distributed *Chlorella* sp. was recognized. From this, it was suggested that, after pesticide dispersion in the river water, the species composition of the microcosm ecosystem came to be more affected by an increase in river water volume.

9.3.6 River Water Containing Pesticides and Herbicides

River water was targeted for evaluation by the raising of a seedling box with insecticide (Gazette granules) and a weedkiller (precutting emulsion, crushing 1 kg of granules, Mamet SM granules) in a rice field after dispersion, before dispersion, and with the passage of water, and it was sampled for a day. Additionally, the rice field consisted of upper reaches, where it is thought that there is an inflow of pesticide, areas in which it is thought that pesticide does not flow into the lower reaches, and areas in which it is thought that there is an inflow of pesticide.

An inorganic salt solution and 5 mg/L of polypeptone Taub's basal medium were added to the river water, which was then filtered and sterilized. An inorganic salt solution and 5 mg/L of polypeptone Taub's basal medium were added to the experimental system and Milli-Q water and then poured by 200 mL. Finally, a TP$_{50}$ nutrient medium was used in the control system, and the microcosm was subcultured in a TP$_{50}$ nutrient medium with a microcosm seed at 2800 lux (L/D = 12 h/12 h), 25 °C, without stirring. The species composition of the microcosm did not exhibit any remarkable changes. When compared with the control system, at the location where it was assumed that there was no inflow of pesticide, no substantial difference was observed before and after weedkiller dispersion.

At the upper reaches of the farm, no change was observed in the abundance of *Chlorella* sp., the threadlike blue-green alga, *Tolypothrix* sp., or in the rotifer, *Philodina erythrophthalma*, in the microcosm using river water sampled on the day when herbicidal dispersion was completed at the farm. However, a decrease was observed in the abundance of the rotifer, *Lecane* sp., the oligochaete, *Aeolosoma hemprichi*, and in a bacterium. Meanwhile, an increase was observed in the abundance of the protozoan, *Cyclidium glaucoma*. Moreover, a decrease in the abundance of *Lecane* sp. was remarkable, and their presence was not able to be confirmed by the 21st day of the experiment.

At a downstream location at the farm, no change was observed in the abundance of *Chlorella* sp., *Tolypothrix* sp., *Philodina erythrophthalma*, or *Aeolosoma hemprichi* in the microcosm using river water sampled on the day when herbicidal dispersion was completed at the farm. However, a decrease was observed in the abundance of *Lecane* sp. and a bacterium, and an increase was observed in the abundance of *Cyclidium glaucoma*. Additionally, the decrease in the abundance of *Lecane* sp. and in bacteria were remarkable, and the presence of *Lecane* sp. was not able to be confirmed by the 21st day of the experiment.

No difference was observed after culturing began in the upper reaches or in the downstream area of the farm until the seventh day, when a difference between these areas became noticeable. A decrease was seen in the downstream abundance of *Lecane* sp. and in a bacterium in comparison with the upper reaches on the 14th day after culturing began, and an increase in abundance was observed in *Cyclidium glaucoma*. Additionally, at the location of inflow, where it was assumed that there was no influence from the dispersion of pesticide because there was no control system, remarkable differences in the species composition of the microcosm were confirmed.

9.3.7 Golf Course Effluent

The investigation of golf course effluent was conducted for underdrainage after pesticide dispersion at a golf course in Narita City, Chiba. The pesticides used were all herbicides, namely, Double Up DG, Scoly Tex (SDS), Ajiran, Mecoprop (MCPP), and Dictran. Additionally, field measurements of water temperature, pH, and DO and laboratory measurements of T-N, NO_2-N, NO_3-N, NH_4-N, T-P, PO_4-P, COD, BOD, and SS were taken. The ratios of the test drainage water were set as 0% (control/no-addition system), 25%, 50%, 75%, and 100%. After 16 days of microcosm cultivation, test drainage water, which was suction filtered using a 0.45 μm diameter membrane (ADVANTEC, GS-45), was added to the microcosm.

As the addition ratio of the environmental water increased to 25%, 50%, and 100%, the structural parameters (abundance) of *Chlorella* sp. (a producer), *Scenedesmus quadricauda* (a producer), and *Aeolosoma hemprichi* (a top predator) were influenced. In the 100% addition system, *Aeolosoma hemprichi* perished. From these results, the m-NOEC of the structural parameter was estimated as less than 25% because some influence was observed at all addition ratios of in the golf course effluent. As shown in Fig. 9.8, system bioactivity increased in the 25% addition system for the functional parameter (P/R ratio); it then decreased before rising again in the 50% addition system. Furthermore, bioactivity decreased remarkably and immediately in the 100% addition system. From these results, the m-NOEC of the functional parameter was estimated as 25% of the golf course effluent. Dilution of more than four times was considered necessary to discharge this drainage safely (Fig. 9.8).

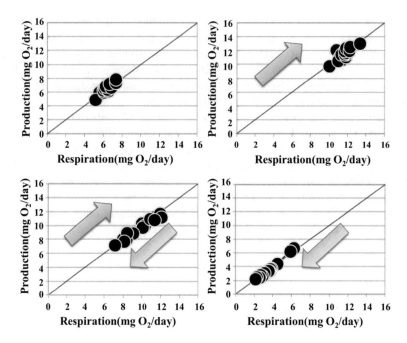

Fig. 9.8 Time course of P/R ratio in golf course effluent added microcosm-WET test
➡Activity of microcosm system increased in time course
➡Activity of microcosm system decreased in time course
*The activity in each concentration was increased in 25% addition with influence to phytoplankton, increased initially and then decreased in 50% addition with influence to both phyto- and zooplankton, and decreased in 100% addition with zooplankton perish

9.3.8 Brewing Plant Effluent

The drainage from an industrial site located in a northwestern area of Chiba Prefecture that was near a brewing factory was tested. Water temperature, pH, and DO were measured in the field, and T-N, NO_2-N, NO_3-N, NH_4-N, T-P, PO_4-P, COD, BOD, and SS were measured in the laboratory. The ratios of the test drainage water were set as 0% (control/no-addition system), 25%, 50%, 75%, and 100%. After 16 days of microcosm cultivation, test drainage water, which was suction filtered using a 0.45 μm diameter membrane (ADVANTEC, GS-45), was added to the microcosm. The species composition on the 30th day (N30) was contrasted with that of the control system. Whether the species abundance ratio to the control system in each microcosm for each microorganisms in the addition systems increased or decreased was evaluated.

As a structural parameter, all constituent species for the microcosm existed in all addition systems. With respect to increases and decreases in the abundance of each microorganism in comparison with the control system, there was no significant

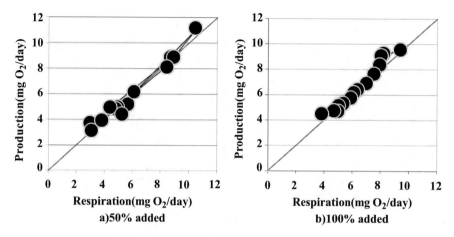

Fig. 9.9 Time course of P/R ratio in brewing plant effluent added microcosm-WET test. (a) 50% added (b) 100% added

change in the population in the 50% addition system. Conversely, *Chlorella* sp. (a producer), *Scenedesmus quadricauda* (a producer), and *Tolypothrix* sp. (a producer) decreased in their abundance, and the abundance of *Philodina erythrophthalma* (a predator) increased in the 100% addition system. From these results, the m-NOEC of the structural parameter was estimated as 50%. Comparison of the DO and P/R ratio of each system revealed that the P/R ratio was close to one in all microcosm systems. The system bioactivity increased and then decreased in the 50% addition system and significantly decreased immediately in the 100% addition system. Consequently, the m-NOEC of the functional parameter was estimated as less than 50% because some influences were recognized in all addition systems. Dilution of more than two times was judged as necessary to discharge this drainage safely (Fig. 9.9).

9.3.9   Metal Processing Plant Effluent

The drainage from an industrial site located in a central area of Chiba Prefecture and near a metal processing factory was tested. Water temperature, pH, and DO were measured in the field, and T-N, NO_2-N, NO_3-N, NH_4-N, T-P, PO_4-P, COD, BOD, and SS were measured in the laboratory. The ratios of the test drainage water were set as 0% (control/no-addition system), 25%, 50%, 75%, and 100%. After 16 days of microcosm cultivation, test drainage water, which was suction filtered using a 0.45 μm diameter membrane (ADVANTEC, GS-45), was added to the microcosm.

For the structural parameter, no significant decrease in the abundance of each microorganism in the microcosm was observed in the 25% addition system when

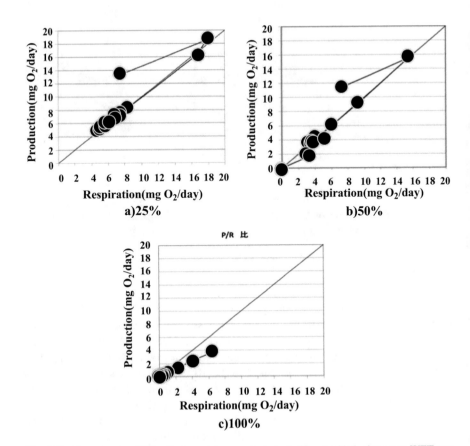

Fig. 9.10 Time course of P/R ratio in metal processing plant effluent added microcosm-WET test. (**a**) 25% (**b**) 50% (**c**) 100%

compared with the 50% and 100% addition systems. The abundance of predatory micro-animals decreased in the 50% addition system, more than in the 25% addition system. The abundance of the producers, *Chlorella* sp., *Scenedesmus quadricauda*, and *Tolypothrix* sp., decreased in the microcosm, and the top predator, *Philodina erythrophthalma*, became extinct in the 100% addition system. From these results, the m-NOEC of the structural parameter was estimated as 25%. As a functional parameter, the P/R ratio was close to 1 in all microcosm systems. In comparison with the control system, the system bioactivity (i.e., the value of DO) of the 25% addition system rose and then decreased before finally stabilizing; that of the 50% addition system rose and then decreased to 0, and that of the 100% addition system decreased to 0 immediately after the addition of drainage water. Put succinctly, the bioactivity of the microcosm increases to maintain itself with low drainage input but cannot rise, and it collapses with high drainage input. From these findings, the m-NOEC of the functional parameter was estimated as 25%. Dilution of more than four times was judged as necessary to discharge this drainage safely (Fig. 9.10).

9.3.10   Landfill Leachate

Jeong et al. (1995, 1996) examined landfill leachate as a very minor pollutant component of drainage and evaluated its influence on *Daphnia magna* in both a single-species culture test and in a multiple-species microcosm test. Landfill leachate (untreated), ozone-treated leachate, fluidized bed-treated leachate, and ozone + fluidized bed-treated leachate were applied to the microcosm-WET test. The ratios of the test drainage waters were set as 0 (control/no-addition system), 5, 10, 25, 50, and 100%. At the beginning of microcosm cultivation, test drainage water, which suction filtered using a 0.45 μm diameter membrane (ADVANTEC, GS-45), was added to microcosm.

As a result of the untreated landfill leachate, the NOEC of the swimming inhibition of *Daphnia magna* was 2.5%, and the half-influence concentration (EC$_{50}$) was 6.0%. The NOEC for the populations of the constituent species in the microcosm test were evaluated for the 1–5% addition of untreated landfill leachate. All species in the microcosm survived and were coexisting on the 28th day in the 1% addition system, and almost no influence of landfill leachate was observed. Meanwhile, for the 5% addition system, microbial species, except for bacteria and *Chlorella* sp., became extinct after the seventh day, and biological interactions among producers, consumers, and decomposers had collapsed. From these outcomes, it was estimated that there exists a boundary concentration for system stability between 1% and 5% of landfill leachate. *Cyclidium glaucoma*, the most sensitive species, exhibited growth similar to that in the control system until the fourth day, and it was clearly influenced by the landfill leachate during the first 7 days.

9.3.11   Ozone-Treated Landfill Leachate

From the addition of ozone-treated landfill leachate, all microbial species except for *Aeolosoma hemprichi* survived and coexisted, and almost no influence was observed in the 1% addition system. Meanwhile, all species except for bacteria and *Chlorella* sp. became extinct after the seventh day, and biological interaction collapsed in systems with more than 5% addition. These results were very similar to those for the addition of untreated landfill leachate. Two interpretations of these results were considered, one in which hazardous materials in landfill leachate still remained and influenced the microcosm in spite of ozone treatment and another in which new hazardous materials were generated by ozone treatment. It has been reported that humic in landfill leachate converts to aldehyde or carboxylic acid, which are toxic mutagens, after ozone treatment. From this, the possibility of generating new hazardous materials may be assumed. Furthermore, it was estimated that there exists a boundary concentration for the stability of the ecosystem between 1% and 5% of ozone-treated landfill leachate. *Cyclidium glaucoma* exhibited similar growth as in the control system until the fourth day, and it was clearly influenced by the ozone-treated landfill leachate during the first 7 days.

9.3.12 Fluidized Bed-Treated Landfill Leachate

From the addition of fluidized bed-treated landfill leachate, all microbial species except for *Lecane* sp. and *Aeolosoma hemprichi* survived and were coexisting on the 28th day in the 100% addition system, and it was estimated that the influence of fluidized bed-treated landfill leachate was minimal because the microbial interactions among producers, consumers, and decomposers were maintained. This tendency did not change in the lower addition systems until 10% addition, and the concentration at which all microbial species, including *Lecane* sp. and *Aeolosoma hemprichi*, could coexist was estimated as less than 10%. This is due to the fact that *Lecane* sp. and *Aeolosoma hemprichi* lack resistance to fluidized bed-treated landfill leachate. In comparison with untreated landfill leachate and ozone-treated landfill leachate, in which almost all microbial species became extinct at concentrations of more than 5%, the influence of fluidized bed-treated landfill leachate on the ecosystem was clearly reduced. Namely, it was clear that landfill leachate that may greatly influence ecosystems can be changed to treated water, which has a lesser influence on the ecosystem through the use of fluidized bed treatment.

9.3.13 Ozone + Fluidized Bed-Treated Landfill Leachate

The results of the addition of ozone + fluidized bed-treated landfill leachate were similar to those from the addition of fluidized bed-treated landfill leachate; all microbial species, except for *Lecane* sp. and *Aeolosoma hemprichi*, survived and were coexisting on the 28th day in the 100% addition system. It was estimated that the influence of ozone + fluidized bed-treated landfill leachate was also minimal because the microbial interactions were maintained, and this tendency did not change in lower addition systems until 10% addition. The concentration at which all microbial species, including *Lecane* sp. and *Aeolosoma hemprichi*, could coexist was estimated as less than 10%. This is due to the fact that neither *Lecane* sp. nor *Aeolosoma hemprichi* are resistant to ozone + fluidized bed-treated landfill leachate. In comparison with the ozone-treated landfill leachate, in which almost all microbial species became extinct at concentrations of more than 5%, the influence of ozone + fluidized bed-treated landfill leachate on the ecosystem was clearly reduced. It is thought that this is because the aldehyde or carboxylic acid, which are by-products of ozone treatment, was decomposed and mineralized by fluidized bed treatment. Put succinctly, it was made clear that the ozone-treated landfill leachate, which greatly influences the ecosystem, can be converted to treated water, with a reduced influence on the ecosystem by combination with fluidized bed treatment. From these results, the microcosm test can be assumed to be representative of the swimming inhibition test of *Daphnia magna*, which is a single-species test regulated by the OECD. Therefore, the utility of the microcosm test as a multiple-species culture system that is a closer approximation of real ecosystems was made clear. After assessing the impacts of landfill leachate from the oxygen demand that followed a 100-day oxygen consumption principle of the landfill leachates, namely, the Ultimate BOD, the NOEC for the 100-day oxygen consumption agreed with the m-NOEC. From this finding, the effectiveness of the microcosm test was supported.

It is important to assess the complex influences of chemical agents, such as surfactants and medicines, on the environment. To conduct such comprehensive assessments, microcosm-WET tests should be used to examine drainage by collecting total toxicity test data through single-species surveys of chemical contaminants in environmental water. It is thought that more details for high-risk materials can be analyzed by the microcosm-WET test with a general comparison analysis. Additionally, because the microcosm-WET test offers high precision at a low cost, and it is expected that the correlation between the natural ecosystem and the microcosm is high, the estimated NOEC is more realistic than that derived from single-species tests, and the prediction of the behavior of the natural ecosystem is more practically enabled. It can also be expected that wastewater management can greatly contribute to the remediation of the environmental impacts of effluents by introducing the microcosm-WET test of the Ministry of the Environment, Japan.

Literature Cited

Japan Sewage Works Association. Wastewater examination method -FY2012-. Tokyo; 2012.
Tatarazako N. The newest domestic and overseas trend of assessment and regulation of effluent water using bioassay–overseas operation case to Japanese WET test introduction. Tokyo: NTS; 2014. 322pp.

Chapter 10
A Scaled-Up Model Ecosystem Verification of the Microcosm N-System

Yuhei Inamori, Ryuhei Inamori, and Yuzuru Kimochi

Abstract In this chapter, a scaled-up microcosm system used for verification of the microcosm N-system is described. The microcosm N-system is recognized as a very useful tool for environmental impact assessment, but it is necessary to verify that it reflects the behavior of natural ecosystems well. To evaluate this, a scaled-up microcosm system was constructed and investigated.

10.1 Overview of the Scaled-Up Model Ecosystem

10.1.1 Species Composition

A scaled-up model ecosystem can be constructed by introducing submerged plants, fish, bivalve mollusks, shrimp, and other organisms into the microcosm N-system, which consists of conventional producers, predators, and decomposers (Takagi et al. 1994; Takamatsu et al. 1995; Inamori and Takamatsu 1995). The model ecosystem with aquatic animals and plants is shown in Fig. 10.1. Representative test organisms in temperate areas include *Egeria densa*, a submerged plant; *Anodonta woodiana*, a freshwater bivalve mollusk; and the fish *Rhodeus ocellatus* and *Pseudorasbora parva*. With the coexistence of these organisms, creating an ecosystem through the combination of submerged plants and aquatic animals, the construction of a scaled-up model for co-culturing large aquatic organisms was conducted. It was then possible to analyze the influence of chemical substances on the selected model ecosystem. The species used in this scaled-up model ecosystem included *Potamogeton malaianus*, *Vallisneria denseserrulata*, *Elodea nuttallii*, *Potamogeton dentatus*, and *Potamogeton pusillus* as producers; the fish *Rhinogobius flumineus*;

Y. Inamori · R. Inamori
Foundation for Advancement of International Science, Tsukuba, Ibaraki, Japan
e-mail: y_inamori@fais.or.jp; r_inamori@fais.or.jp

Y. Kimochi (✉)
Center for Environmental Science in Saitama, Kazo, Saitama, Japan
e-mail: kimochi.yuzuru@pref.saitama.lg.jp

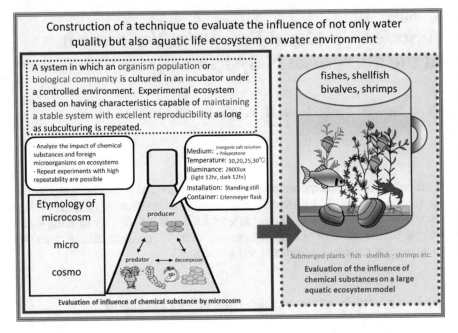

Fig. 10.1 Overview of scaled-up model ecosystem

and the arthropod predator *Caridina multidentata*. A stable model ecosystem hydrosphere can be constructed using the appropriate combination of species.

10.1.2 Culture Vessel and Culture Conditions

An example of an ecosystem model using aquatic animals and plants is shown in Fig. 10.2. The experimental apparatus should contain glass beads (1 L) for planting submerged plants and 4 L of lake water in a 5 L beaker. The apparatus should be installed in a thermostatic chamber at 20 °C that blocks light from the outside. An irradiance of 5000 lux and fluorescent lamps that can be adjusted to create day and night (L/D) cycles should be used.

10.2 Chemical Substance Addition Experiment

10.2.1 LAS

An experiment was performed for the purpose of constructing a standard by planning the construction of the model ecosystem (scaled-up model ecosystem) hydrosphere using larger aquatic organisms in addition to the microcosm N-system, evaluating

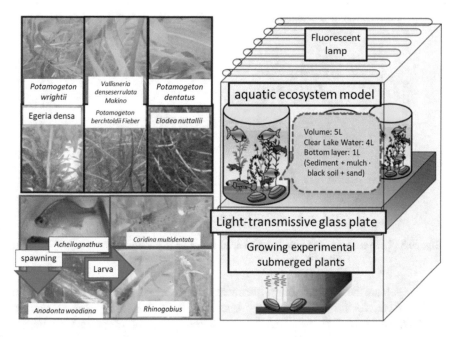

Fig. 10.2 Water plants and animals supplied to scaled-up model ecosystem and its culture system

the risk associated with the introduction of chemical substances in the model ecosystem. To construct an optimum, scaled-up model hydrosphere ecosystem, which consisted of plants, fish, bivalve mollusks, and crustaceans, the abundance input organisms, submerged plants, feed quantity, and bottom sediment quality were selected. Additionally, experiments were conducted to determine whether it is possible to maintain the number of initial input organisms with a constant abundance of individuals under the 24-h light condition without setting an L/D cycle. As a result, there were periods in which input organisms could survive continuously for more than 1 week, but high mortality indicates that the experimental aquatic ecosystem was far from stable. Therefore, when the abundance of organisms varied due to mortality, an experiment for evaluating the number of viable organisms was conducted without introducing new organisms. As a result, there were no fatalities for 10 days for three *Rhodeus ocellatus*, two *Rhinogobius* sp., four *Caridina multidentata*, and two *Anodonta woodiana* at the time that the powdered potato feed was supplied under the experimental condition with the highest survivorship. We determined abundance to be a viable input. Furthermore, the viable abundance under the L/D = 12 hr./12 hr. condition that was similar to natural conditions was evaluated, and the model ecosystem hydrosphere in the 5 L plastic tank (i.e., the scaled-up model ecosystem) included three *Rhodeus ocellatus*, one *Rhinogobius* sp., three *Caridina multidentata*, and two *Anodonta woodiana* based on laboratory findings. *Vallisneria denseserrulata* was planted, supplied with powder feed, and subjected to filling by pollutants + humus and black soil + sand.

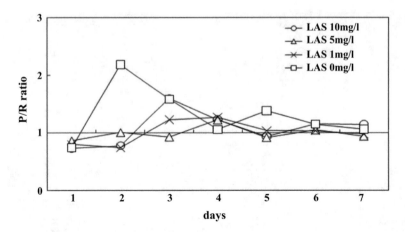

Fig. 10.3 Time course of P/R ratio in LAS added scaled-up model ecosystem

To evaluate the influence of the chemical substance LAS on the model ecosystem hydrosphere, LAS was added to the system at 0, 1, 5, and 10 mg/L. Based on fluctuations in the P/R ratio and the persistence of aquatic organisms, an impact assessment of the model ecosystem was conducted. As shown in Fig. 10.3, it was assumed that the functional parameter (P/R ratio) in the experiment with the powder feed supply, which fluctuated rapidly only when 10 mg/L was added, and some other effects were latent. Also, in the structural parameters (i.e., persistence of aquatic animals), from the first day to the third day after the addition of 10 mg/L, two *Rhodeus ocellatus*, one *Rhinogobius* sp., and one *Anodonta woodiana* died. The risk at 10 mg/L became obvious, and the deaths of one *Anodonta woodiana* on the first day at 5 mg/L and one *Rhodeus ocellatus* on the seventh day at 1 mg/L were also confirmed. The results of the P/R ratio of the experiment without the supply of powder feed exhibited no change with the addition of any amount of LAS, and it was determined that there was no influence. With respect to the persistence of the aquatic animals, one *Rhodeus ocellatus* each died on the first and second day with the 10 mg/ L addition and with the 1 mg/L addition, and one *Anodonta woodiana* and one *Caridina multidentata* died on the second and third day, respectively.

Catabolite suppression was considered to have possibly occurred due to the difference between experiments with a feed supply and those without a feed supply. With catabolite restraint, rapid metabolism of organic matter is a phenomenon known to restrain the generation of the enzyme necessary for stable organic metabolism. Powder feed quantity and the inspection of the metabolic rate of LAS are necessary to determine whether feeding resulted in more sudden changes in the P/R ratio than when there was no feed supplied. Additionally, as a result of changing the quality of the glass beads on the bottom and having performed a similar experiment, it was determined that, when the P/R ratio and the mortality of aquatic organisms were assessed, there was a risk for LAS = 10 mg/L. Plasticity was required, but it was made clear that the NOEC of the LAS estimated was less than 5 mg/L.

10.2.2   AE

A scaled-up microcosm as a model aquatic ecosystem, including aquatic animals and plants, which was nearer to the conditions of a real ecosystem than was the microcosm N-system was constructed, and the ecosystem risk posed by a nonionic surfactant, alcohol ethoxylate (AE), as a chemical substance was assessed. As a result of examining the appropriate combination of large aquatic animals and plants, a stable, large-scale model ecosystem was constructed as the ecological impact assessment system in which *Rhodeus ocellatus* and *Rhinogobius* sp. (aquatic fish) and *Egeria densa* (a submerged plant) coexisted as consumers and a producer, and the P/R ratio was approximately constant at 1. Additionally, it may be useful to construct a stable ecosystem index using the P/R ratio. The NOEC of AE in this scaled-up aquatic microcosm was estimated as 2 mg/L. As a result of AE addition to the scaled-up microcosm, aquatic animals were not affected up to 2 mg/L of AE, and the submerged plant was not affected up to 5 mg/L of AE. The influence of AE was felt in the scaled-up microcosm at additions of 3–5 mg/L, and the P/R ratio increased above 1, exhibiting similar behavior to the system that included submerged plants only.

Using consecutive measurements of the DO in the scaled-up microcosm, ecological balance and a steady state can be informed, and the persistence of aquatic populations can be judged. An activity state can be quickly derived from the changing pattern (inspection of the recovery accuracy) of the DO level or the P/R ratio. The possibility of evaluating ecological risk from the DO pattern and the P/R ratio, without measuring the population of the aquatic organisms and using a continual measurement of DO, was illustrated. The NOEC of this scaled-up microcosm was nearly equal to the m-NOEC of the N-system, which was correlated with the flask microcosm test as shown in Fig. 10.4. This indicates that, even if the diversity and hierarchical structure of the organisms change, there is no substantial difference in the influence concentration. More specifically, the validity and effectiveness of the microcosm N-system was demonstrated by the experimental results of the scaled-up microcosm system (Fig. 10.4).

10.3   Comparison with the Microcosm N-System

In the reports of Sugiura (Sugiura 2009, 2010), it was presumed that, in a microcosm where the production and respiration can be balanced, even if the diversity of the constituent species and the hierarchical structure are changed somewhat, there is no substantial difference in the influence of the concentration. The correspondence

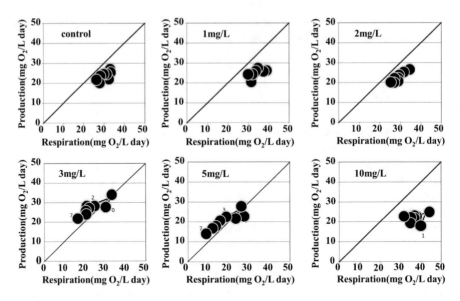

Fig. 10.4 Time course of P/R ratio in AE added scaled-up model ecosystem

between the NOEC in the flask-scale microcosm and the large-scale ecosystem model, with larger aquatic animals and plants, suggests that it is possible to predict the difference in the influence of the concentration across different ecosystem scales and species compositions by collecting basic data. Furthermore, in the large-scale ecosystem model, it is possible to evaluate not only the fish (predators) but also the influence of aquatic plants (producers) on ecosystem functions. It is a great advantage that the survivorship/mortality of the aquatic animal populations can be quickly determined from the changes in the DO value or the P/R ratio. By collecting such basic data, it is possible to ensure the accuracy of the correlation between these model ecosystems and natural ecosystems. As a result, it is expected that the NOEC of chemical substances can be accurately predicted for natural ecosystems using large-scale models of aquatic ecosystem. It is also expected to be useful as a tool for managing chemical substances, including surfactants.

Based on the results of experiments with the addition of LAS and AE, the ecosystem function parameters (DO and P/R ratio) in the scaled-up model ecosystem were consistent with the microcosm N-system. Therefore, the microcosm N-system adequately reflects the natural ecosystem, and it was shown that it is extremely useful as a model of ecosystem functions. In the OECD test method, one representative species from each trophic level of the ecosystem was selected (e.g., *Selenastrum capricornutum* (producer), *Ceriodaphnia dubia* (primary consumer), *Danio rerio* (high-order consumer), bacterial luminescence inhibition test (decomposer), etc.) to evaluate toxicity. The ecosystem effect is then calculated based on the toxicity data for the most sensitive species. In natural ecosystems, biological interactions, material circulation, and energy flows exist and differ greatly from the physiological activities of test organisms in single-species tests. The evaluation of toxicity under different conditions of physiological activity is inadequate, especially for impact assessments of ecosystem function; thus, the necessity of multi-species tests is noted at the OECD. Meanwhile, mesocosm tests, in which natural environmental water is sealed in a container or in which part of a natural ecosystem is isolated, suffer from problems in terms of cost, reproducibility, and handling. The mesocosm test seems to be meaningful in predicting the behavior of chemical substances evaluated in the presence or absence of higher predators in the ecosystem. However, the test method in which fish and other organisms are added is considered to not be useful because it is inferior to the flask microcosm in terms of cost, speed, and reliability. Considering these points, it can be said that the particle size of the microcosm N-system is extremely useful as a multi-species coexisting system, with high reproducibility and stability.

Literature Cited

Beyers RJ, Odum HT. Ecological microcosms. New York: Springer; 1993. 557pp.

Graney RL, Kennedy JH, Rodjers JH Jr. Aquatic mesocosm studies in ecological risk assessment. Boca Raton: Lewis Publishers; 1994. 723pp.

Peterson JE, Kennedy VS, Dennison WC, Michael Kemp W. Enclosed experimental ecosystems and scale; tools for understanding and managing coastal ecosystems. London: Springer; 2009. 221pp.

Matsuda H. Ecological risk science for beginners: precautionary adaptive management. Tokyo: Kyoritsu Shuppan; 2008. 213pp.

Saijo Y, Sakamoto M. Experimental analysis of lake ecosystem using mesocosms. Nagoya: The Nagoya University Press; 1993. 346pp.

Chapter 11
A Subsystem Microcosm Verification of the Microcosm N-System

Yuhei Inamori, Ryuhei Inamori, and Kazuhito Murakami

Abstract Various impact assessments are enabled using the microcosm N-system as described in the previous chapters, and the consistency of the microcosm with the natural ecosystem was demonstrated experimentally. However, analysis of the structural parameter (abundance) becomes particularly complicated in comparison with the single-species test prescribed by the OECD. Therefore, a microcosm subsystem was constructed, which combines one species each of microorganisms as a producer, consumer, and decomposer. Together, these were investigated for the possibility of using this subsystem for impact assessment (i.e., it was inspected with respect to its superiority relative to the N-system).

11.1 Outline of the Microcosm Subsystem

11.1.1 Species Composition

The microcosm subsystem was constructed with the bacterium *Pseudomonas putida* as a decomposer, the protozoan *Cyclidium glaucoma* as a consumer, and the chlorophyte *Chlorella vulgaris* as a producer. *Pseudomonas putida* and *Cyclidium glaucoma* were isolated from the microcosm N-system, and *Chlorella vulgaris* was isolated from the Microorganisms Culture Center of the NIES, Japan. Another microcosm subsystem composed of the euglenid *Euglena gracilis* as a producer, the ciliate *Tetrahymena thermophila* as a consumer, and the bacterium *Escherichia coli* as a decomposer was also constructed from a pure strain culture. All microorganisms in these microcosm subsystems can be observed ordinally in natural ecosystems. Each simplified system exhibited high reproducibility and stability, similar to the microcosm N-system.

Y. Inamori (✉) · R. Inamori
Foundation for Advancement of International Science, Tsukuba, Ibaraki, Japan
e-mail: y_inamori@fais.or.jp; r_inamori@fais.or.jp

K. Murakami
Chiba Institute of Technology, Narashino, Chiba, Japan
e-mail: kazuhito.murakami@p.chibakoudai.jp

Y. Inamori (ed.), *Microcosm Manual for Environmental Impact Risk Assessment*,
https://doi.org/10.1007/978-981-13-6798-4_11

11.1.2 Culture Vessel and Culture Conditions

The culture vessel and culture conditions were similar to those used in the microcosm N-system. A 300 mL Erlenmeyer flask was used, and the conditions included a temperature of 25 °C and an illuminance of 2400 lux, with an L/D cycle of 12 h each.

11.2 Mathematical Simulation

Differential equations were developed to simulate the dynamics of a batch culture microcosm that was composed of the bacterial decomposer, *Pseudomonad putida*; the protozoan consumer, *Cyclidium glaucoma*; and the chlorophyte producer, *Chlorella vulgaris*. Two different types of equations were examined. One was an equation that accounts for the metabolites of microbes, and the other was an equation that does not account for them. The results of the simulation were compared with the empirical data obtained from the substrate-one species and substrate-two species subsystems and from the microcosm N-system. It was shown that the differential equations with the variables for the metabolite and a lack of protozoan predation on bacteria corresponded better with the empirical data than did the equations without these variables. Therefore, it was suggested that the promotion and inhibition of microbial growth by each metabolite and the existence of non-predated bacteria by protozoans were important to the dynamics of a batch culture microcosm.

The population dynamics of genetically engineered microorganisms (GEMs) in a microcosm were also analyzed using computer simulations. The GEM *Escherichia coli* HB101/pBR325 model was inoculated into the microcosm containing the alga, *Chlorella vulgaris*; the bacterium, *Pseudomonas putida*; and the protozoan, *Cyclidium glaucoma*, 14 days after cultivation of the microcosm was initiated. When the model GEMs were added at 10^8 cells/mL (high density), protozoans multiplied and then decreased rapidly. The same population density was recovered as that in the microcosm without the addition of the model GEM. The model GEM also rapidly decreased until it leveled out at a density of $\sim 10^4$ cells/mL when protozoan abundance increased. Conversely, in the microcosm into which model GEMs were added at 10^4 cells/mL (low density), the population density of neither the indigenous bacteria nor the model GEM changed. Based on empirical data, a differential equation that represents the indigenous bacterial and model GEM population dynamics was developed and solved. The simulation corresponded well with the empirical data. However, there was a difference between the high- and low-density introductions of the model GEM in the various parameters, representing the exchange of bacteria between the predated model GEM and the non-predated model GEM by protozoans. This indicates that the values of parameters changed when the model GEM aggregated or when a cellular morphological change of the model GEM occurred.

11.3   Flux Analysis by ^{14}C Addition

In recent years, global warming caused by anthropogenic increases in CO_2 has become a critical problem, and it should be noted that aquatic ecosystems participate deeply as places of CO_2 sources and sinks. However, elucidating the mechanisms of this participation is difficult because natural ecosystems are comprised of a great variety of organisms, with complicated interactions existing between many of them, and various environmental parameters are also related in a complex way. Therefore, to control the environmental parameters, experimental ecosystems using model ecosystems, which are simplified so as not to affect the behavior of the system, have become essential. Here, radioactive carbon (^{14}C) was added to the microcosm subsystem containing *Chlorella vulgaris* (producer), *Cyclidium glaucoma* (consumer), and *Pseudomonas putida* (decomposer), and the rate of carbon flux was examined by measuring the ^{14}C quantity taken in by each constituent microbe. The compartmentalization of the constituent microbes of the microcosm subsystem was fractionated by *Cyclidium glaucoma* + *Chlorella vulgaris* and *Pseudomonas putida* + dissolved matter with a 2.0 μm diameter filter, *Cyclidium glaucoma* and *Chlorella vulgaris* + *Pseudomonas putida* + dissolved matter with a 5.0 μm diameter filter, and microbes + dissolved matter with a 0.2 μm diameter filter.

The RI of the constituent microbes in the microcosm was examined, and the carbon flux after the addition of $NaH^{14}CO_3$ was calculated over time. The uptake of carbon of each microbe was determined using the rate of increase in the RI of each microbe until 30 minutes after chemical addition. As a result, the RIs were 2227 Bq/pop/hr. in *Chlorella vulgaris*, 254.25 Bq/pop/hr. in *Pseudomonas putida*, and 45.019 Bq/pop/hr. in *Cyclidium glaucoma*. Therefore, when *Chlorella vulgaris* took in carbon as a carbonate or carbonate ion, *Pseudomonas putida* took in the products of metabolism, and *Cyclidium glaucoma* took in carbon from prey, the RI was calculated as 1 for *Chlorella vulgaris*, 0.1 for *Pseudomonas putida*, and 0.02 for *Cyclidium glaucoma*; the flux decreased by approximately 1/10 whenever it passed through each trophic level in the microcosm food chain. The value of 1/10 corresponds to 10 times the formula of energy flow in the natural ecosystem, and it was determined that the microcosm can serve as a functional model of natural ecosystems.

11.4   Phosphate Addition

The effects of phosphorus concentrations on aquatic microcosms were evaluated in terms of microbial population dynamics and the photosynthetic activity, which is the starting point of material flux. Microcosms that consisted of *Chlorella vulgaris*, *Pseudomonas putida*, and *Cyclidium glaucoma* were incubated using media that contained various concentrations of KH_2PO_4. At 14–16 days after incubation began, [^{14}C]-sodium bicarbonate was added to each microcosm, and then the microbial uptake of the radioactive substrate was monitored. The population densities of *Chlorella vulgaris* and *Cyclidium glaucoma* increased, as the initial phosphorus concentration increased, while that of *Pseudomonas putida* remained constant. The photosynthetic activity of *Chlorella vulgaris* also increased as the initial phosphorus concentration increased. In the microcosm, *Pseudomonas putida* uses the metabolites produced by *Chlorella vulgaris* and is preyed on by *Cyclidium glaucoma*. Therefore, the growth activity of *Pseudomonas putida* increased as the photosynthetic activity of *Chlorella vulgaris* increased, but predation pressure by *Cyclidium glaucoma* also increased. Consequently, the population of *Pseudomonas putida* remained constant. These findings led to the conclusion that the phosphorus concentration strongly affects microbial population dynamics and the material flux in the microbial loop of the aquatic ecosystem. However, the mass balance of each organism should remain in a steady state in environments where the population of predators, such as protozoans, invertebrates, and fish, are suitable.

11.5 Metal Addition

The complexity (e.g., species diversity) of the system required for environmental impact assessment using the microcosm is an important problem. It becomes difficult to maintain the plasticity of the experiment while the accuracy of the provided toxicity data increases, and it is generally assumed that the cost increases because it more closely approximates the complexity of real (natural) ecosystems. Conversely, in the flask-sized aquatic microcosm, constant toxicity data were provided regardless of the species diversity. Here, Cu, Mn, and Al were loaded on a microcosm subsystem comprised of *Escherichia coli* as a decomposer, *Euglena gracilis* as a producer, and *Tetrahymena thermophila* as a consumer, and their influence on production and respiration rates and the P/R ratio (functional parameter) was compared with the result from the more complicated microcosm N-system, and the superiority was inspected.

In the control group, an irregular daily change in DO, such as increased respiration, was observed until the fifth day after cultivation began, suggesting that the system changed into one that was different from the original microcosm. Such a phenomenon was not observed in the microcosm N-system. Because a pure culture was used for the constituent species of the microcosm subsystem, in contrast to that used in the microcosm N-system that was derived from nature, the subsystem lacked any natural resistance to invasion. This is thought to result in the defeat of various germs in the naturally derived microcosm N-system and the spread of germs, which attached to oxygen electrodes, in the subsystem.

No significant influence was observed with respect to the cell count, whereas the influence on production and respiration rates increased relative to the control system and was observed at ~0.64 mg/L of Cu. Only a slight influence was recognized in the cell count, while the value of each functional parameter decreased at 55 mg/L of Mn more than in the control system. Therefore, functional parameters appear to be more susceptible to chemical substances than to structural parameters.

When Cu was added to the microcosm N-system, the production and respiration rates rose in a concentration-dependent manner between 0.3 and 1.2 mg/L but declined at 1.5 mg/L. The production rate increased at 0.4–1.2 mg/L in comparison with the control system and decreased at 1.5 mg/L. The respiration rate increased at 1.2 mg/L and 1.5 mg/L, higher than in the control system. The P/R ratio was not clearly influenced at 1.2 mg/L or less but decreased at 1.5 mg/L more clearly than in the control system. These effects corresponded to the risk assessment pattern "A-2" (i.e., maintenance of the system, with no influence on the P/R ratio and wherein organismic activity increased) at low concentrations (0.3–1.2 mg/L) and to the "D-2" pattern (i.e., collapse of the system in which the P/R ratio decreased and did not recover) at the high concentration (1.5 mg/L). When Cu was added to the microcosm N-system, there was no apparent influence on the population at 0.6 mg/L or less, and this result is similar to that in the microcosm subsystem. A decrease or increase in the abundance of individual species was recognized in comparison with the control system when 0.9 mg/L or 1.2 mg/L of Cu was added to the microcosm N-system, but no species perished.

When Mn was added to the microcosm N-system, neither functional parameter was affected, nor did they exhibit the "A-1" (i.e., maintenance of the system, with no influence on the P/R ratio and biological activity) pattern at 0.1 mg/L. By adding 1 mg/L of Mn, the production rate was not clearly influenced, and the respiration rate increased only temporarily after the addition of more than that of the control system. As a result, the P/R ratio temporarily decreased after the addition of Mn to less than that of the control system. Therefore, it can be assigned to the risk assessment pattern "B-2" (i.e., recovery of the system in which P/R decreased and recovered). In contrast, by loading the experiment for the microcosm subsystem, 55 mg/L and 550 mg/L of Mn was added and exhibited a "D-2" pattern (i.e., collapse of the system in which P/R decreased and did not recover). When Mn was added to the microcosm N-system, there was no apparent influence of the population observed at 0.1 mg/L. At 1 mg/L, some species perished, and the populations of others decreased or increased in comparison with the control system. However, no species perished even at 5.5 mg/L of Mn added to the microcosm subsystem, and there was little influence on the cell count. Therefore, it was suggested that the sensitivity of structural parameters to Mn was lower in the subsystem than in the microcosm N-system.

It was made clear that the sensitivities of the microcosm subsystem and microcosm N-system were similar or that of the subsystem was lower in both the functional parameter and structural parameter endpoints. However, there was a difference in the addition concentration of loading experiments, and there remains the problem that the loadings only accounted for two kinds of metals. For a stable evaluation, the superiority of the biologically diverse N-system becomes clear.

11.6   Indigenous Bacterial Addition

Using the microcosm subsystem, a combination of *Chlorella vulgaris* (producer), *Cyclidium glaucoma* (consumer), *Pseudomonas putida* (decomposer), the microbial pesticide, *Bacillus thuringiensis* subsp. *aizawai* KH as a nonindigenous bacterium, and *Escherichia coli* HB101/pBR325 as a non-transmissible plasmid-containing indigenous bacterium was added to the microcosm after culturing began, with 10^8 cells/mL on the 16th day (during the stationary phase). Both *Bacillus thuringiensis* subsp. *aizawai* KH and *Escherichia coli* HB101/pBR325 were preyed upon by the protozoan *Cyclidium glaucoma* after addition and decreased to a population of 10^5 cells/mL (after addition the 14th day) by the 30th day. With respect to the predation characteristics of *Cyclidium glaucoma*, a similar result was provided in the predator-prey interaction test, and it was revealed that both nonindigenous bacteria were suitable food sources. The surviving population came to have more than ten times the abundance of the microcosm N-system and was comprised of a greater diversity of biota on the 30th day (after addition on the 14th day). This indicates that the ecosystem is composed of complex interactions and that the interactions become more complicated with the introduction of nonindigenous organisms. Additionally, after mathematical simulation of the *Escherichia coli* HB101/pBR325 addition system, it was revealed that laboratory findings and simulation results were in good agreement with one another by treating the bacteria that were not preyed upon as a different variable than those that were preyed upon. The effects of bacterial cohesiveness and shape were large and greatly accounted for the differences between the predation properties of *Cyclidium glaucoma* at both high and low concentrations of *Escherichia coli* HB101/pBR325.

11.7   Irradiation

The influence of continuous gamma ray irradiation on the microcosm subsystem comprised of the bacterial decomposer, *Escherichia coli*; the ciliate predator, *Tetrahymena thermophila*; and the flagellate producer, *Euglena gracilis*, was evaluated. In the control system, the abundance values of *Escherichia coli*, *Euglena gracilis*, and *Tetrahymena thermophila* were constant at 10^6, 3×10^5, and 10^3 cells/mL, respectively, during the stationary phase. In contrast, the cell count of *Escherichia coli* decreased when they were irradiated for 56 days over more than one column, with both 10 Gy/day and 23 Gy/day, and some *Tetrahymena thermophila* also perished. No influence was observed for *Euglena gracilis* at 10 Gy/day, but there was a slight decrease in their abundance at 23 Gy/day. In contrast, in the microcosm N-system, only bacteria decreased at 10 Gy/day, and bacteria and *Lecane* sp. decreased, and *Tolypothrix* sp. increased at 23 Gy/day. Therefore, it may be said that the microcosm N-system exhibited greater resistance to continuous gamma radiation in comparison with the microcosm subsystem. Because these systems completely differ in the composition of their microorganisms, the predictability of this finding should be verified. However, the microcosm N-system, in which there is a greater diversity of constituent species, may have greater resistance to this stressor.

In the microcosm subsystem, which consisted of three species, a stable coexistence was provided as in the microcosm N-system, which was comprised of 11 species, and thus an impact assessment, such as that of various chemical substances, could reasonably be performed. However, problems arise in the microcosm subsystem when using a single species each of producers, consumers, and decomposers, such as the lack of complex interactions between species and competition or high-level trophic dynamics, which are among the most basic and important elements constituting an ecosystem. Because the microcosm was a model of ecosystem function and regulated to evaluate the DO (a functional parameter) for the endpoint of the impact assessment and the P/R ratio for the complexity of the operation, it was assumed that a more detailed environmental assessment was enabled in the microcosm subsystem by utilizing positioning for a complementary subsystem of the microcosm N-system. In other words, the microcosm subsystem can be used, but greater accuracy and detail are provided using the microcosm N-system, which includes more diverse biota. From such a point of view, the superiority of the microcosm N-system for risk evaluations is clear.

Literature Cited

Jorgensen SE. Modeling for fate and effectof toxic substances in the environment. In: Developments in environmental Modelling 6. Amsterdam: Elsevier; 1984. 342pp.

Landner L. Chemicals in the aquatic environment, advanced hazard assessment. London: Springer; 1989. 415pp.

Leffler JW. The use of self-selected, generic aquatic microcosms for pollution effects assessment. In: White HH, editor. Concepts in marine pollution measurement. College Park: University of Maryland; 1984. p.139–58.

Resetatits WJ Jr, Bernoado J. Experimental ecology, issues and perspectives. Oxford: Oxford University Press; 1998. 470pp.

Chapter 12
Further Perspectives

Ryuhei Inamori, Yuhei Inamori, Kazuhito Murakami, and Kaiqin Xu

12.1 Analysis and Assessment of the Microcosm N-System for OECD Standardization

Positioning of the various docimasy of the bioassay and the fundamental utility of the microcosm for ecosystem impact assessments are shown in Figs. 12.1 and 12.2. In addition to various factor analysis at the ecosystem level using the microcosm N-system, the combination of the general interaction analysis through the use of a simulated environmental microcosm (mesocosm) and the underlying interactions analysis through both the two-species cultures and the single-species cultures, are important for the advancement of ecosystem impact assessment. It is the microcosm test that becomes essential for evaluating ecosystem impacts, including those on the human body, based on the study of marmots. Additionally, although the docimasy varies in various bioassays, the microcosm test remains at the core.

A procedure for assessing the impacts of chemical substances by culturing, data collection, and branching-type analyses of variance using the microcosm N-system docimasy is shown in Fig. 12.3. Specific examples of the evaluation in this docimasy are shown below.

1. Environmental assessment of 2,3,4,6-tetrachloro phenol (TCP), an insecticide, sterilizer, wood preservative, and formicide. The most susceptible creature, water

R. Inamori (✉) · Y. Inamori
Foundation for Advancement of International Science, Tsukuba, Ibaraki, Japan
e-mail: r_inamori@fais.or.jp; y_inamori@fais.or.jp

K. Murakami
Chiba Institute of Technology, Narashino, Chiba, Japan
e-mail: kazuhito.murakami@p.chibakoudai.jp

K. Xu
National Institute for Environmental Studies, Tsukuba, Ibaraki, Japan
e-mail: joexu@nies.or.jp

© Springer Nature Singapore Pte Ltd. 2020
Y. Inamori (ed.), *Microcosm Manual for Environmental Impact Risk Assessment*,
https://doi.org/10.1007/978-981-13-6798-4_12

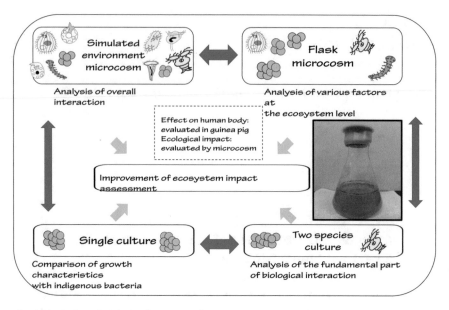

Fig. 12.1 Basic principle for optimizing of environmental impact risk assessment

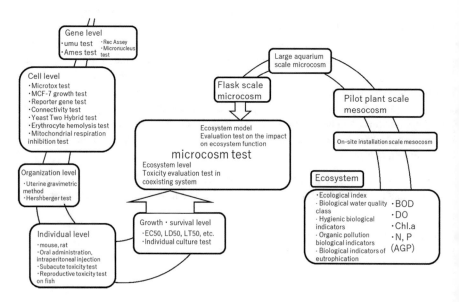

Fig. 12.2 Placement of bioassay test and microcosm test

fleas; ECOSAR class, phenol. The amount of production and respiration decreased even with the addition of 1 mg/L (Fig. 12.4). The amount at which the influence on production became 0 was 10 mg/L, and it was strongly indicated

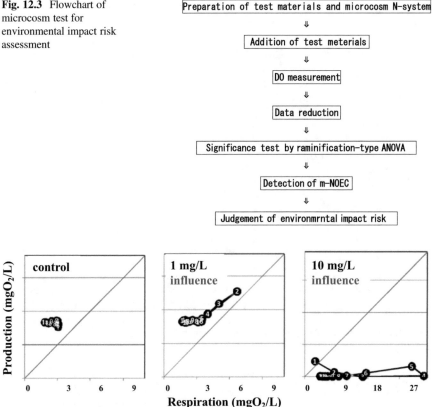

Fig. 12.3 Flowchart of microcosm test for environmental impact risk assessment

Fig. 12.4 Time course of P/R ratio in TCP added microcosm

that TCP had a greater influence on production than on consumption. As a result of having let match document investigations, a similar tendency was shown in the mesocosm examination, and there was a close association found in the final examination.

2. Environmental assessment of carbendazim, a fungicide. The most susceptible creature, water fleas; ECOSAR classes, imidazoles and carbamate esters. In both systems, it was at approximately one that the P/R ratio stably fluctuated, but, after recovery following attenuation in the 10 mg/L addition system, the amplitude increased with the addition of 20 mg/L (Fig. 12.5). The m-NOEC of the Benrate hydration agent was estimated as ~20 mg/L. Because a Benrate hydration agent is the main product of carbendazim and is composed primarily of ~50% benomyl, and the benomyl is converted into carbendazim with a 2-hour half-life in water, the m-NOEC of the carbendazim is thought to be ~10 mg/L.

3. Environmental assessment of chlorpyrifos, an insecticide. The most susceptible creature, water fleas; ECOSAR classes, esters and monothiophosphates. There

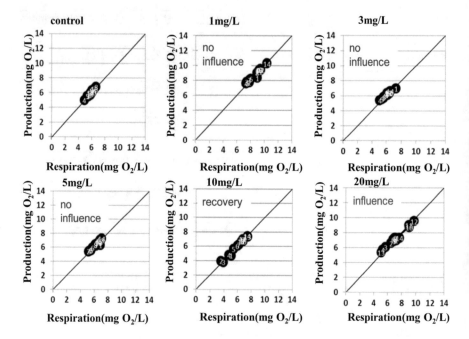

Fig. 12.5 Time course of P/R ratio in carbendazim added microcosm

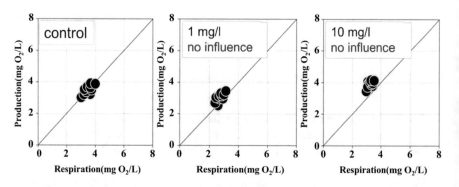

Fig. 12.6 Time course of P/R ratio in chlorpyrifos added microcosm

was no influence on the production (P) or respiration (R) in either concentration zone (Fig. 12.6). However, chlorpyrifos has notable hydrolysis and soil adsorption characteristics, and intermittent administration is performed in outdoor dispersion experiments (i.e., in natural ecosystem experiments/mesocosms). It is thought that the difference between the experimental conditions is related to the presence or absence of its influence.

4. Environmental assessment of alachlor, a weed killer. The most susceptible creature, algae; ECOSAR class, haloacetamides. There was no statistically

Fig. 12.7 Time course of
P/R ratio in alachlor added
microcosm

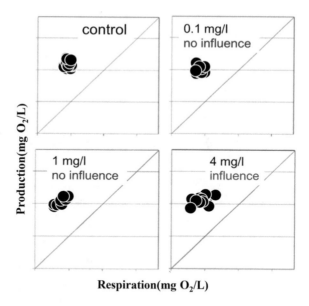

significant difference at concentrations of 0.1 mg/L or 1 mg/L, though significant
differences were detected at 4 mg/L (Fig. 12.7). The m-NOEC of the alachlor was
1–4 mg/L.
5. Environmental assessment of linuron, a weed killer. The most susceptible creature, algae; ECOSAR class, substituted urea. It was observed that a nearly stable
   state was reached at a concentration of 0.1 mg/L, but the amount of production
   and respiration decreased in all of the systems with more than 0.5 mg/L
   (Fig. 12.8). Thus, the m-NOEC of linuron was estimated as 0.1 mg/L in the
   impact assessment based on the P/R ratio.

The results of the impact assessment of P and R data by a ramification-type
ANOVA using MS Excel revealed that there was no influence at addition concentrations of 1 mg/L of TCP, 20 mg/L of carbendazim, 4 mg/L of alachlor, or 0.5 mg/L
of linuron. There was also no influence with either addition concentration of
chlorpyrifos.

Chlorpyrifos, which deviated from the confidence interval from the correlation
between the NOEC of the microcosm N-system test and the NOAEC of the outdoor
ecosystem test (mesocosm test), exhibited substantial soil adsorption. Additionally,
paraquat, which is a weed killer, also deviated from the correlation of this test and the
outdoor experiment but can protect nature ecosystems because it is above the value
divided by the safety factor (200). Because chlorpyrifos is less toxic in the
mesocosm experiment without soil, and the microcosm N-system test also does
not have soil, it was estimated to be less toxic in the microcosm test as well.
Therefore, it is necessary to consider the influence of the degradability of the test
material and the soil.

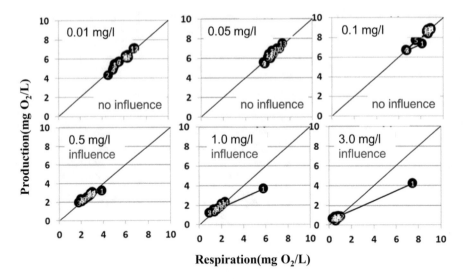

Fig. 12.8 Time course of P/R ratio in linuron added microcosm

Based on the results of these evaluation procedures, widespread chemical sub-stances were grouped based on the characteristics of each chemical substance. The accumulation and inspection of practical ecosystem impact data can support and render possible the OECD international standardization of the microcosm test. When considering the past flow of OECD standardization, the following are needed: (1) confirmed reproducibility by blind testing of the same materials by three insti-tutions, (2) inspection of the versatility using the ring test performed by five domestic institutions, (3) construction of a practical infrastructure by compiling the m-NOEC database, (4) making the general examination methods accessible by completion of the test manual, and (5) inspection of the practicality of the microcosm N-system by developing a NOEC prediction method.

For steps 1–4, the reinforcement, careful investigation, and efficiency of micro-cosm experiment should be improved, and the microcosm test manual should be accuracy based on the characteristics and addition time of the chemical substances to microcosm, that is, at a steady state or at culture starts. With these procedures, the generalization and practical realization of the microcosm test is possible. For step 5, the database of m-NOEC of the microcosm test and the NOECeco of the mesocosm test will be expanded, the applicable range of the microcosm will be established, and the unevenness of each chemical substance from the correlation formula, the difference in the sensitivity of each ecosystem, and the rationale behind the assessment factor have been already established. Furthermore, the limits of the microcosm test suggested in this study may be clarified based on the evaluable characteristics of chemical substances. It has already been clarified whether this technique can be used to extrapolate to natural ecosystems. In other words, the utilization of this microcosm test and strengthening of the manual are made possible by the expansion of the database, which is the basis for the analyses shown in this manual and which is necessary for expanding the applicable range and optimization of the test method. A close inspection of the assessment factor reinforced in this manual has already been completed.

In the future, the microcosm test will be established as a practical ecosystem impact assessment technique, replacing the mesocosm test by further grouping chemical substances after having thoroughly investigated their characteristics. Because it is anticipated that the correlation between natural ecosystems and microcosms is high, it is thought that the estimated NOEC is more realistic than that derived from evaluations using single-species. It currently remains difficult to omit the test methods for algae, crustaceans, and fish represented by the WET test in the ecosystem impact assessments of chemical substances. However, by using a parallel ecosystem model that includes a food chain and energy flow, it becomes possible to accumulate knowledge, such as that related to the degradability and residual properties of chemical substances, and the recovery and collapse of ecosystem functions. In other words, it may be said that the superiority of the microcosm is enhanced by quantifying and evaluating ecosystem impacts. The WET test is a docimasy for evaluating toxicity that includes combined influences by testing water containing multiple chemical substances, but it is a single-species test that does not allow for the evaluation of individual chemical substances. Although it uses species in different niches in the food chain for an assay, it must be noted that the WET test has a problem when testing under conditions in which there are interactions among material circulations, energy flow, and biological interaction, which are the foundations of natural ecosystems. The microcosm, which is a system of multiple, coexisting species, evaluates the risk of chemical substances at an ecosystem level, and the safety factor is obtained from the results of research on the correlation between the conventional mesocosm test and microcosm test before the PNOEC can be calculated from the accumulation of further data. The merit to the regulations of the chemical industry is remarkable.

The need for evaluating the introduction of chemical substances in model ecosystems has been noted by the OECD and other chemical substance management entities, and regulatory laws and model systems are being developed. However, an international standard for ecosystem impact assessments using mesocosms and microcosms has yet to be undertaken. The methods described in this book allow for the influence of chemical substances on ecosystems to be evaluated in the microcosm, and the effects on the human body, by using marmots and mice, were demonstrated using the microcosm N-system. It is possible to determine that there is no risk to the ecosystem by the docimasy, analysis, and evaluation of the NOEC described in this book. Because a standardized microcosm test could be constructed, this approach should be utilized globally.

12.2   Summary and Further Perspectives

Although ecological risk assessment has been conducted on the risk posed by toxic chemicals to an individual species, the microcosm method can assess the state of recovery and change of an ecosystem by fragmentation, which is not shown in single-species tests. Further, the microcosm test can be used to conduct assessments equivalent to those of conventional mesocosms, measuring changes in the abundance of organisms. The scientific importance of being able to continuously measure changes in the function of an ecosystem is great. This study suggests that even when a microcosm experiences structural effects, the functional stability of the ecosystem can be maintained due to the redundancies at the population or community level. This shows that when a functional effect is emphasized in an ecosystem impact assessment, proper assessment cannot be conducted by measuring structural parameters (species composition, abundance, etc.), which are generally measured, and that it is necessary to measure functional parameters directly.

The microcosm test method was shown as effective for assessing the impacts of biological interactions, which might be overlooked in single-species tests. In particular, there were protozoans and metazoans in the microcosm, which have not conventionally been used, and it was indicated that the existence of these taxa was important for evaluating the influence on the microbial ecosystem. Additionally, it was clarified that the P/R ratio assessment method using the model ecosystem hydrosphere (microcosm) could acquire the basic information needed to determine the influence of various chemical substances on the ecosystem and that it was a useful technique for protecting natural ecosystems using the appropriate management of chemical substances. As for this assessment, an experimental method and data analysis techniques are made available in the manual. Moreover, this ecosystem impact assessment technique has become frequently used in Japan, and it offers the possibility for OECD standardization. It may be said that the significance of this approach for international environmental policy is extremely high. Additionally, the importance of the patterns and environment impact assessment was adopted in the final report of a task group of the Association of National Irradiation Ecology, titled "Ecosystem approach to environmental protection," and this report may possibly have a substantial influence on the International Commission on Radiological Protection (ICRP) or the Organization for Economic Cooperation and Development Nuclear Energy Agency (OECD/NEA). It is expected that the microcosm test method will contribute to an environmental radioprotection policy in Japan through the advice of these international organizations, and it is thought that the ripple effect could be tremendous. Additionally, it is at an early stage of development by the lead laboratory and the domestic ring test in Japan, and it is the establishment and unification of the test method (protocol), inspection (ring test) of the plasticity between facilities, construction of the database, and the clarification of the application range that promote its prospects as an OECD standardized test method with future development of the microcosm test. At this stage, it is necessary to improve the database, inspect correlative analyses with natural ecosystems (mesocosms), and evaluate the relationship with known toxic data. Having passed through these tasks, it will be established as a public, fixed method and could have wide utility after validation by other institutions and peer review (international examination). Furthermore, by following the WET test method and utilizing the results provided by this project to their maximum, it is expected this technique will provide new environmental policy information from Japan for development as an international, standardized OECD test.

In the optimized microcosm, producers, predators, and decomposers are properly structured, and they can be used for highly accurate environmental impact assessments. It can be said that the optimized microcosm test system is not the current OECD test method for individual organisms, such as *Daphnia* or algae, utilized for the WET test, but a test system that approximates natural ecosystems as closely as possible. Therefore, more realistic results at the ecosystem level are expected to be obtained by using the microcosm test method. In order to determine the standard of treated water quality in wastewater treatment, it is important to consider how the trace substances in the treated water affect the aquatic ecosystem at the discharge destination. Also, in promoting environmental impact assessment, it is necessary to evaluate various chemical substances that may affect aquatic ecosystems. According to the OECD test method, representative species are selected one by one from each trophic level of the ecosystem (e.g., *Selenastrum* (producer), *Ceriodaphnia* (primary predator), *Danio rerio* (higher predator), luminescent bacteria (decomposers), etc.), toxicity assessments are then performed, and then the ecosystem impact is calculated based on the toxicity data for the most sensitive species. However, in natural ecosystems, the biological activity of test organisms is different from that of single-species tests because there are biological interactions, material circulation, and energy flows in natural ecosystems. Therefore, toxicity assessments under different biological active conditions are insufficient, particularly as impact assessments of ecosystem function. For this reason, it has been noted that the OECD should conduct various biological tests. It is very important to develop an evaluation tool from aquatic ecosystem models (Takamatsu et al. 1995). To evaluate the toxicity of chemical substances to ecosystems with complex biological interactions, microcosms of model ecosystems composed of producers (algae), predators (microanimals), and decomposers (bacteria) are useful tools. Whole aquatic ecosystems are composed of food webs (i.e., biological interactions, including high-order predators, such as fish), based on microbial loops that are mainly composed of microbes. To evaluate the effect of chemical loads on complex ecosystems, it is important to consider multiple parameters, including the production (P)/respiration (R) ratio. Establishing risk management methods for ecosystems based on environmental changes and the analyzed relationships between the microbial organisms and higher predators that constitute aquatic ecosystems becomes very important (Fig. 10.1). For ecosystem impact assessments, it is effective to comprehensively utilize the kinetic analysis of microbial communities using the microcosm and other multi-species mixed model ecosystems (stable model ecosystems).

This research was conducted to establish a standard test method characterized by both high plasticity and low cost using the microcosm that has been stably subcultured for more than 40 years. The basic manual that affected the testing operations was already built (Ministry of the Environment, Environment study synthesis promotion costs: FY2009–FY2011), and it was frequently used to perform tests between facilities. The OECD examination guidelines progressed through the ring tests, and the establishment of an OECD standard docimasy was planned. The outcomes obtained from this investigation are listed below.

1. Screening test of chemical agents using the microcosm

The experimental test was advanced by screening 198 listed chemical substances into three groups, those that acted on animals, plants and algae, or bacteria, according to differences in their mechanisms. As a result, the foundational information of an OECD test was provided because the microcosm test did not depend on the classification of chemical substances as mentioned previously, the sensitivity was higher than in a single-species test, and it was strongly correlated with the mesocosm test.

2. Environmental impact assessment of chemical agents using the microcosm

For establishment as a practical ecosystem impact statement technique for the mesocosm test method, the examination of widespread chemical substances was promoted. A positive ion surfactant, TMAC (1000 mg/L); an environmental hormone, nonylphenol (100 mg/L); a sterilizer, mancozeb (300 mg/L), the weed killers, alachlor (1000 mg/L), linuron (0.1 mg/L), and fomesafen (30 mg/L); and an insecticide, chlorpyrifos (500 µg/L), which were provided for microcosm examination (m-NOEC), were matched with published data, and most of the 26 substances examined exhibited similar behaviors to those in the field experiment (mesocosm test). It became clear that there was a strong association between the microcosm and the mesocosm.

3. Analysis of the relationship between the microcosm test and the mesocosm test

As described in Chap. 8, the validity of the toxicity evaluation by the microcosm test was considered, and it was shown that 23 of 26 substances adhered to the confidence interval of the regression line and the strength of the correlations among the microcosm, experimental ecosystem (mesocosm), and the natural ecosystem was demonstrated by the means of the m-NOEC and the NOEC of the natural ecosystem. The lower limit of the confidence interval is also a straight line obtained by dividing the mean of the natural ecosystem NOEC by the uncertainty coefficient (coefficient of assessment = 200) calculated in consideration of the differences in ecosystem sensitivity. Only the lower limit line is required to determine the PNOEC of a natural ecosystem, and, if the point on the correlation plot is greater than this (i.e., above the lower limit line), then it is measured in the microcosm (PNOEC) obtained by dividing the NOEC by the coefficient of assessment (200). With a value that is lower than that of the natural ecosystem (mesocosm test) PNOEC, it can protect the natural ecosystem. Most chemicals measured in this investigation fit into this range (i.e., above the lower limit line). On the other hand, while substances exceeding the upper limit can predict the PNOEC, the value is less than in natural ecosystem NOECs. Only 1 substance out of 26 fell above the upper limit line, revealing that precision is high in the microcosm test, regardless of the chemical substance in question.

4. Comparison and interface of the microcosm and mesocosm

As described in Chap. 8, the two substances that deviated from the confidence interval of the correlation between the mean NOEC of the natural ecosystem and the m-NOEC are characterized by high soil adsorption. Based on the results of this research, it is necessary for most environmental impact assessments of chemical substances to show the limits of the microcosm and plan for OECD international standardization. The outcomes of this investigation are shown on the next page as a poster presented on August 28, 2014 (study briefing session on LRI). The need for developing evaluation techniques for the introduction of chemical substances into model ecosystems has been noted by chemical substance management organizations, including the OECD, and regulatory laws and model systems are being established. However, international standardization of the ecosystem impact statement technique using the mesocosm and microcosm has not yet been completed. The standardization of the microcosm test method for environmental impact assessment should be promoted strongly and as soon as possible.

Appendices

Appendix 1: Complementary Assessment Methodology

Kazuhito Murakami and Ken-ichi Shibata
Chiba Institute of Technology, Narashino, Chiba, Japan
Yokohama National University, Yokohama, Kanagawa, Japan

Abstract Impact assessment using the microcosm should be evaluated based on the P/R ratio, namely, a functional parameter. However, it can be evaluated from the P/R ratio using structural parameters, such as population measurements. Although the analytical procedure described in this book outlines the basic principles for environmental assessments of chemicals using the microcosm N-system, the analytical methods needed for understanding the microcosm are described here. A method of impact assessment was used that was based on the respiration amount (functional parameter) as an endpoint other than the biomass (population) and production (P)/respiration (R) ratio (functional parameter). Additionally, the culture period and philosophical underpinnings of the influence of this approach are described.

K. Murakami
Chiba Institute of Technology, Narashino, Chiba, Japan
e-mail: kazuhito.murakami@p.chibakoudai.jp

K. Shibata
Yokohama National University, Yokohama, Kanagawa, Japan
e-mail: shibata-kenichi-ym@ynu.jp

© Springer Nature Singapore Pte Ltd. 2020
Y. Inamori (ed.), *Microcosm Manual for Environmental Impact Risk Assessment*,
https://doi.org/10.1007/978-981-13-6798-4

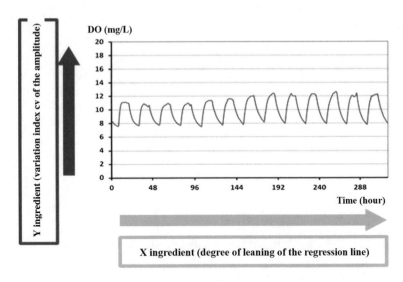

Fig. 1 Separation of dissolved oxygen (DO) waveform into x gradient and y gradient

Simplified Statistical Process

All microcosms, including the control system, are treated as equivalent, and, for the microcosm (N-system), it is assumed that the difference between the flask ($\pm 1\sigma$ both in structure and function) can be ignored due to its stability and reproducibility, which has extended for more than 40 years. The microcosm was cultured in triplicate for each experimental condition, and the obtained data were evaluated using a Student's *t*-test for significance. The coefficients of variation (i.e., the standard deviation and mean, expressed as percentages) for the production rate and the respiration rate in the microcosm from the beginning of the culture to the 7th day were 20% or less (N = 25 independent experiments), and those from the 14th day to the 21st day were 5% or less. These values are nearly identical to previously reported values.

At first, the change in DO should be decomposed into an *x*-component (i.e., the slope *a* of the regression line) and a *y*-component (i.e., the coefficient of variation *cv* of the amplitude) (Fig. 1).

- *x*-component (slope of the regression line)...Find a regression line from a DO graph using Microsoft Excel to observe the trajectory (i.e., transition of the system) of the DO, and calculate the slope of each addition concentration, such that:

$$y = ax + b; \qquad a : \text{slope}; \quad b : y\text{-intercept}$$

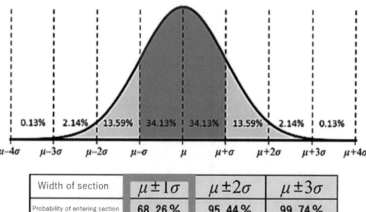

Fig. 2 Standard division and probability

- *y-component* (coefficient of variation *cv* of the amplitude)...Find the coefficient of variation *cv* from the mean and standard deviation of the DO to determine the amplitude (i.e., the activity of the system) of the DO.

$$cv = \sigma/\mu \qquad \sigma : \text{standard deviation} \quad \mu : \text{population mean}$$

When the microcosm is cultured three (n = 3) or five (n = 5) times, find the mean slope *a* and the coefficient of variation *cv* for each sample size (n). Then, suppose that each numerical value (i.e., the means) of the slope *a* and the coefficient of variation *cv* is normally distributed. In a normal distribution, as shown in Fig. 2, the probability of settling is within ± 1σ of the mean (i.e., of falling within one standard deviation and can be considered as an error), which is 68.26%. Therefore, evaluations should not be influenced if the slope *a* and the coefficient of variation *cv* in each additional system are within ± 34.13% of each value in the control system. Additionally, in order to consider whether ± 1σ represents an error, (i) assume that the error in the control system fits within ± 1σ, and (ii) evaluate the influence more strictly.

In addition, the basis of ± 1σ is (1) supposing the error in the control system fits within range of ± 1σ and (2) evaluating it more severely.

Respiration Amount Assessment

The quantity of respiration is, as shown in Fig. 3, the geometrical expression of the total amount of respiration within an area of the graph, which is displayed over an

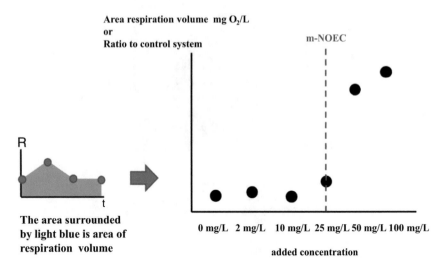

Fig. 3 Example for m-NOEC estimation by area respiration quantity

elapsed period of time on the vertical axis after the addition of a chemical substance for the horizontal axis in the amount of respiration, R (from the 16th day to the 30th day), until the culture end time.

The amount of respiration is determined as the total for all active organisms (microalgal producers, microanimals (consumers), and bacterial decomposers) comprising the microcosm. Calculate the amount of respiration in every additional concentration in the microcosm test, and estimate the maximum concentration at which a meaningful difference is not recognized within the control system (no-addition system) as m-NOEC (microcosm no observation effect concentration).

Population Assessment

Here, the evaluation of population change in the structural parameter is described. Using the methods described in Chap. 6 (Sect. 6.1.), measure the biomass (population) for a day, and transcribe the measurement of the biomass to the graph of the population (N/mL or cm/mL) on a vertical axis and the cultivation time (day) on the horizontal axis. Using the graph, perform the following (1–3) steps.

(1) Evaluate the behavioral patterns after the addition of the chemical substance. Evaluate the behavioral patterns for each constituent microbe for each 14-day addition, in comparison with the control system (no-addition system).
(2) Evaluate the population at the time that the culture ends (N_{30}). Evaluate the population of each constituent microbe (after addition on the 14th day) in

comparison with the control system (no-addition system) 30 days after the culture began.

(3) Evaluate the abundance of constituent organisms after the addition of the chemical substance (B_{16-30}). Evaluate the abundances of constituent microbes for each 14-day addition, in comparison with the control system (non-addition system). Calculate abundance graphically using the following equation (1), with the quantity of existence (area population density):

$$\text{abundance} = \sum [\{(\text{population } N_1 + \text{population } N_2)/2 \times \text{days (days)}\}] \quad (1)$$

Additionally, a more detailed evaluation and analysis is enabled by comparing the functional parameters (DO and P/R ratio) with the structural parameters (abundance and biomass).

Influence and Assessment Method

As for the period after adding a chemical substance to the microcosm, the acute influence (16–20 days) may be defined as B_{16-20}, while the subacute influence (20–23 days) may be defined as B_{20-23}, and the chronic influence (23–30 days) may be defined as B_{23-30}. The influence on the ecosystem can be evaluated by the different mechanisms of the chemical substance (Fig. 4).

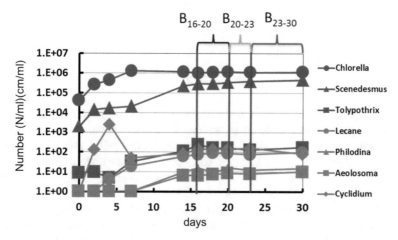

Fig. 4 Acute, subacute, and chronic influence on microcosm test

Relative Community Metabolism(RCM)

To investigate the ecological impacts of a widely used antibiotic, oxytetracycline (OTC), on the population abundance of composite species, gross primary productivity (GPP) and community respiration (CR) were considered. The population abundance, GPP, and CR were linked to understand the relationship among these measures. First, the experimental respiration rate of heterotrophs (HE_{exp}) was obtained from CR less algal respiration calculated from $0.35 \times$ GPP. Next, the population respiration of each heterotroph was calculated from the direct measure of population abundance and the assumed constant of *per capita* respiration. The sum of the population respirations across all heterotrophs was defined as the theoretical respiration rate of heterotrophs (HR_{theo}). Finally, the relative community metabolism (RCM) of the heterotrophic community was obtained by calculating the ratio of HR_{exp} to HR_{theo}, which indicates changes in the specific respiration rates as a whole within the community. The influence of OTC on the RCM was greater than on CR, and thus the effect of OTC on the metabolism of heterotrophs was far more severe than expected from CR. The rate of change in the original data can be magnified by RCM, which facilitates the detection of influences in ecotoxicological studies (Shibata et al. 2014).

The concentration of DO in the culture medium was continuously measured with a handheld optical DO meter (ProODO; YSI/Nanotec Inc., Kawasaki, Japan). The oxygen flux at the atmosphere/medium interface was corrected according to Odum (1956). The respiration under light, which could not be measured directly, was estimated by assuming it to be equivalent to the respiration in the dark. The amount of oxygen consumed during the dark period (12 h) was doubled to indicate the oxygen consumption during an entire 24-h period and defined as the CR. The oxygen consumed during the dark period was combined with the amount of oxygen produced during the light period and defined as the GPP (McConnel 1962). Algal respiration (AR) was obtained from Duarte and Cebrian (1996) using the equation $AR = 0.35 \times$ GPP. The HR_{exp} was calculated as CR less AR:

$$HR_{exp} = CR - AR.$$

The respiration rate of each heterotrophic organism was estimated using body size and population abundance. Respiration of the bacterial community, dominated by *Bacillus cereus*, was estimated by the mean metabolic rate of *Firmicutes* (Makarieva et al. 2012); the bacterial CFU count was assumed as being equal to the cell count. The SR of *Cyclidium glaucoma* was calculated from the cell volume using the equation of Fenchel and Finlay (1983). The cell volume was calculated from their cell length and width measurements (n = 20) using an ellipsoid formula. The SRs of *Lecane* sp., *Philodina erythrophthalma*, and *Aeolosoma hemprichi* were calculated from their dry weights using the equation of Galkovskaya (1995). The dry weights of *Lepadella patella* and *Rotaria rotatoria* + *Philodina roseola* Ehrbg reported by Dumont et al. (1975) were used for the dry weights of *Lecane* sp. and

Philodina erythrophthalma, respectively. The dry weight of *Aeolosoma hemprichi* was estimated from length and width measurements (n = 20) according to the method of Andrassy (1956) as cited by McIntyre (1969):

$$\text{dry weight} = 0.25 \times 1.13 \times \text{length} \times \text{width } 2/1.7,$$

in which the dry weight is determined in grams, and length and width are determined in centimeters. The population-level respiration rate was calculated from the population abundance and the SR. The sum of the population-level respiration rates across all heterotrophic organisms was defined as the HR_{theo}:

$$HR_{theo} = \Sigma(SR_i \times N_i),$$

where SR_i is the SR of a heterotrophic organism i, and N_i is the population abundance of the organism i.

The SRs that are included in HR_{theo} are constant respiration rates based on the following reference data: (1) mean metabolic rate for bacteria (Makarieva et al. 2012), (2) mean metabolic rate between starving and growing conditions for protozoa (Fenchel and Finlay 1983), and (3) "regular metabolism" at typical levels of activity for rotifers (Galkovskaya 1995). This means that HR_{theo} is based on the assumption that the SRs may vary. Thus, the ratio of HR_{exp} and HR_{theo} represents a metabolic rate of the community relative to the average metabolic rate. Here, this is defined as the RCM, such that:

RCM = (actual community metabolic rate)/(reference community metabolic rate).

The RCM can be applied to various groups of organisms (e.g., RCM of microbes, rotifers, fishes, algae, etc.). Here, HR_{exp} and HR_{theo} were used as an actual community metabolic rate and a reference community metabolic rate, respectively. Then the RCM of heterotrophs was calculated as:

$$RCM = HR_{exp}/HR_{theo}.$$

For more information, refer to the work of Shibata et al. (2014) (Fig. 5).

Pattern of Prosperity and Decay of Microorganisms

The populations of the constituent organisms of the microcosm N-system (control system) targeted for evaluation are A (N/ml), B (N/ml), and C (N/ml), as shown in Fig. 6. When populations of the constituent organisms (e.g., 2 weeks later) are assumed, D (N/ml), E (N/ml), and F (N/ml) are considered for a fixed period of time during which chemicals were added to the microcosm: if A ≒ D, B ≒ E, and

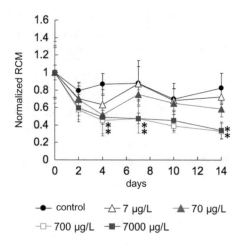

Fig. 5 Time course of RCM after OTC addition to microcosm N-system

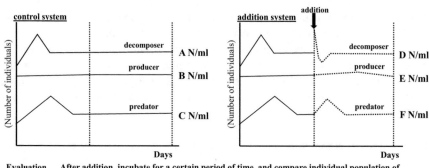

Evaluation criteria After addition, incubate for a certain period of time, and compare individual population of each constituting microorganism with control system

$A \doteqdot D$
$B \doteqdot E$ no effect
$C \doteqdot F$

$A \neq D$
$B \neq E$ There is influence if either one is applicable
$C \neq F$

The number of individuals should be counted at least two points at the time of addition (for example, 16 days after the start of culture) and after a certain period of culture (for example, 16 days after addition), and the change in useful microorganisms after addition is tracked as necessary.

Fig. 6 Assessment from effect on indigenous microorganisms in microcosm

$C \doteqdot F$, there is no influence; if in even one instance $A \neq D$, $B \neq E$, $C \neq F$, there is an influence that may be attributed to the addition of the chemical substances.

Additionally, a condition shown in Fig. 7 is considered with respect to the behavioral patterns when alien species are introduced into the microcosm. Of these, a decreasing trend (Fig. 7a, b) is determined if there is no influence.

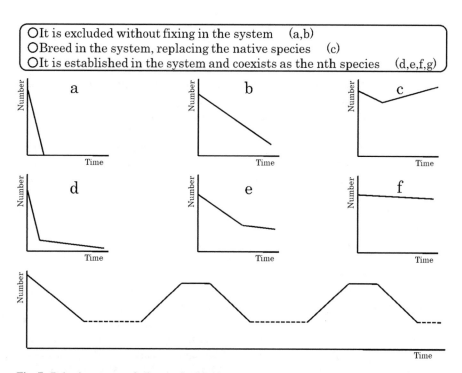

Fig. 7 Behavior pattern of alien species in microcosm

Appendix 2: Suggested Readings

Investigations of microcosms as model ecosystems have been conducted by many researchers across many regions and countries. Here, published, representative microcosm studies are listed.

Research Books of Microcosm Study

1. Kurihara, Y.:Limited Ecology-System of stable and coexistence, *Iwanami Shinsho*, 1975, 187pp. (in Japanese)
2. Kawabata, Z.:Structure Change Analysis of Plankton Population using Microcosm, Microbial Ecology 15 -Ecological model and its utilization, Japanese Society of Microbial Ecology, ed., *Japan Society Press*, 1987 (in Japanese)
3. Kurihara, Y.:Ecology of Coexistence, *Iwanami Shinsho*, 1998, 239pp. (in Japanese)
4. Kurihara, Y.:Ecological System and Human, *Tohoku University Press, Sendai*, 2008, 136pp. (in Japanese)
5. Inamori, Y., Fuma, S. and Inamori, R.:Environmental Assessment of Aquatic Ecosystem on Nuclear Power Plant Accident, Pollution Control Measures of Radioactive Materials after The Great East Japan Earthquake, *NTS*, 2012 (in Japanese)

Research Papers of Microcosm Study

1. Sugiura, K., Sato, S., Goto, M.: Effects of β-BHC on Aquatic Microcosm, *Chemosphere*, Vol.37, No.1, pp.39–44 (1976)
2. Kurihara, Y.: Studies of Succession in a Microcosm, *Science Report of Tohoku University, 4th Ser. (Biology)*, No.37, pp151–160 (1978)
3. Kurihara, Y.: Studies of the Interaction in a Microcosm, *Science Report of Tohoku University, 4th Ser. (Biology)*, No.37, pp.161–177 (1978)
4. Kawabata, Z., Kurihara, Y.: Computer Simulation Study on the Relationships between the Total System and Subsystems in the Early Stages of Succession of the Aquatic Microcosm, *Science Report of Tohoku University, 4th Ser. (Biology)*, No.37, pp.179–204 (1978)
5. Kawabata, Z., Kurihara, Y.: Computer Simulation Study on the Nature of the Steady State of the Aquatic Microcosm, *Science Report of Tohoku University, 4th Ser. (Biology)*, No.37, pp.205–218 (1978)
6. Sugiura, K.:Effect of Cu^{2+} Stress on an Aquatic Microcosm: A Holistic Study, *Environmental Research*, Vol.27, pp.307–315 (1982)
7. Shikano, S., Kurihara, Y.: Community Responses to Organic Loading in a Microcosm, *Japanese Journal of Ecology*, Vol.35, No.3, pp.297–305 (1985)
8. Sugiura, K., Aoki, M., Kaneko, S., Daisaki, I., Komatsu, Y., Shibuya, H., Suzuki, H, Goto, M.:Fate of Trichlorophenol, Pentachlorophenol, p-Chlorobiphenyl, and Hexachlorobenzene in an Outdoor Experimental Pond: Comparison between Observations and Predictions based on Laboratory Data, *Archives on Environmental Contamination and Toxicology*, Vol.13, pp.745–758 (1984)
9. Sugiura, K.:Testing for Effects of Chemicals on Lake and Pond Ecosystems, *Japan Journal of Water Pollution Research*, Vol.8, No.10, pp.175–184 (1985) (in Japanese)
10. Shikano, S., Kurihara, Y: Analysis of Factors Controlling Responses of an Aquatic Microcosm to Organic Loading, *Hydrobiologia*, Vol.169, pp.251–257 (1988)
11. Kawabata, Z.: Use of Microcosms to Evaluate the Effect of Genetically Engineered Microorganisms on Ecosystems, *Proceedings of Advanced Marine Technology Conference*, Vol.3, pp.41–48 (1990)
12. Kurihara, Y.:Stability of Microbial Community, *Science Journal*, Vol.61, pp.611–618 (1991)
13. Shikano, S., Luckinbill, L. S., Kurihara, Y.: Changes of Traits in a Bacterial Population associated with Protozoal Predation, *Microbial Ecology*, Vol.20 (1990)
14. Kurihara, Y., Shikano, S., Toda, M.: Trade-off between Interspecific Competitive Ability and Growth Rate in Bacteria, *Ecology*, Vol.71, No.2 (1990)
15. Inamori, Y. and Sudo, R.: Environmental Assessment of Foreign Microorganisms using Microcosm System, *Kaiyo monthly*, Vol.23, No.1, pp.18–26 (1991) (in Japanese)

16. Sugiura, K.: Effects of Chemicals on an Aquatic Microcosm: Can a Simple Microcosm be Used as a Tool for Generic Ecosystem-level Toxicity Screening?, *Journal of Sagami Women's University*, Vol.55, pp.1–12 (1991) (in Japanese)

17. Inamori, Y., Murakami, K., Sudo, R., Kurihara, Y. and Tanaka, N.: Environmental Assessment Method for Field Release of Genetically Engineered Microorganisms using Microcosm Systems, *Water Science and Technology*, Vol.26, No.9–11, pp.2161–2164 (1992)

18. K.Murakami, Y.Inamori, R.Sudo, Y. Kurihara: Effect of Temperature on Prosperity and Decay of Genetically Engineered Microorganisms in a Microcosm System, *Water Science and Technology*, Vol.26, No.9–11, pp.2165–2168 (1992)

19. Sugiura, K.: A Multispecies Laboratory Microcosm for Screening Ecotoxicological Impacts of Chemicals, *Environmental Toxicology and Chemistry*, Vol.11, pp.1217–1226 (1992)

20. Sugiura, K.: Toxicity Assessment using an Aquatic Microcosm: Effects of Acid Precipitation, *Journal of Sagami Women's University*, Vol.57, pp.5–10 (1993) (in Japanese)

21. Tanaka, N., Inamori, Y., Murakami, K., Akamatsu, T. and Kurihara, Y.: Effect of Species Composition on Stability and Reproductivity of Small-scale Microcosm System, *Water Science and Technology*, Vol.30, No.10, pp.125–131 (1994)

22. Takagi, H., Hashimoto, M., Takamatsu, Y., Inamori, Y.: Assessment of Effect of Herbicides on Aquatic Ecosystem using Small-scale Microcosm Systems, *Journal of Japan Society on Water Environment*, Vol.17, pp.650–660 (1994) (in Japanese)

23. Sugiura, K.: Laboratory Tests for Chemical Effects on Aquatic Ecosystem Properties, *Journal of Sagami Women's University*, Vol.58, pp.5–10 (1994) (in Japanese)

24. Sugiura, K.: Ecotoxicity Test for Aquatic Ecosystem, *Laboratory Animal Technology and Science*, Vol.7, No.1, pp.1–12 (1995) (in Japanese)

25. Tanaka, N., Inamori, Y., Kawabata, Z., Mori, T., Itayama, T. and Sudo, R.: Simulation Model of Microbial Interaction Taking Account of Metabolites in Batch Culture Microcosm, *Journal of japan Society on Water Environment*, Vol.18, No.2, pp.109–120 (1995) (in Japanese)

26. Tanaka, N., Inamori, Y., Kawabata, Z., Mori, T., Itayama, T. and Sudo, R.: Simulation of the Population dynamics of a Genetically Engineered Microorganism in a Microcosm, *Japanese Journal of Water Treatment Biology*, Vol.31, No.1, pp.33–41 (1995) (in Japanese)

27. Inamori, Y., Takamatsu, Y.: Effectiveness of Aquatic Microcosm System containing Material Cycle as Method of Toxicity and Ecological Assessment, *Journal of Japan Society on Water Environment*, Vol.18, pp.612–617 (1995) (in Japanese)

28. Takamatsu, Y., Inamori, Y., Matsumura, M., Sudo, R.: Environmental Assessment of Surfactant using Aquatic Microcosm System, *Journal of Japan Society on Water Environment*, Vol.18, pp.784–793 (1995) (in Japanese)

29. Jeong, M.,Inamori, Y. and Sudo, R.: Environmental Assessment of Landfill Leachate on Aquatic Ecosystem using Microcosm, *Papers on The Biologically New Technology Symposium*, Vol.3, pp.133–136 (1995) (in Japanese)

30. Sugiura, K. : The Use of an Aquatic Microcosm for Pollution Effects Assessment, *Water Research*, Vo.l.30, No.8, pp.1801–1812 (1995)

31. Sugiura, K.: Toxicity Assessment using an Aquatic Microcosm, *Journal of Sagami Women's University*, Vol.59, pp.57–69 (1995) (in Japanese)

32. Takamatsu, Y., Nishimura, O., Inamori, Y., Sudo, R. and Matsumura, M.: Effect of Temperature on Biodegradability of Surfactants in Aquatic Microcosm System, *Water Science and Technology*, Vol.34, No.7–8, pp.61–68 (1996)

33. Kawabata, Z.: Analysis of Aquatic Ecosystem using Microcosm, *Journal of Japan Society on Water Environment*, Vol.19, No.8, pp.610–614 (1996) (in Japanese)

34. Ishikawa, Y., Anderson, G., Poynter, J., MacCallum, T., Frye, R., Kawasaki, Y., Koike, J., Kobayashi, K., Mizutani, H., Sugiura, K., Ijiri, K., Ishikawa, Y., Saito, T., Shiraishi, A.: Mini ecosystem -Preliminary Experiment on Board STS-77-, *Biological Sciences in Space*, Vol.10, No.2, pp.105–111 (1996)

35. Ishikawa, Y., Shoji, N., Sugiura, K.: A Life Simulation of Colony Formation in Microcosm, *CELSS Journal*, Vol.9, pp.5–12 (1996) (in Japanese)

36. Takamatsu, Y., Inamori, Y., Sudo, R., Kurihara, Y. and Matsumura, M.: Ecological Effects Assessment of Anionic Surfactant on Aquatic Ecosystem Using Microcosm System, *Journal of Japan Society on Water Environment*, Vol.11, No.7, pp.710–715 (1997) (in Japanese)

37. Takamatsu, Y. Inamori, Y., Nishimae, H., Ebisuno, T., Sudo, R. and Matsumura, M.: Environmental Assessment of LAS on Aquatic Ecosystem using Scale-up Microcosm System, *Water Science and Technology*, Vol.36, No.12, pp.207–214 (1997)

38. Sugiura, K.: Changes in Constituent biological Species in a Materially Closed Microbial Ecosystem, *Journal of Sagami Women's University*, Vol.61、pp.125–127 (1997) (in Japanese)

39. Inamori, Y., Takamatsu, Y. and Sudo, R.: Ecological Assessment by Aquatic Microcosm System, *Journal of the Japan Society of Waste Management Experts*, Vol.9, No.5, pp.368–378 (1998) (in Japanese)

40. Nakajima, T., Inamori, Y., Endo, G., Kawabata, Z. and Kurihara, Y.: Fate of foreign Species in Microbial Communities: Ecological Mechanisms of Coexistence and Competitive Exclusion, and Experimental Analysis, *Microbes and Environment*, Vol.13, No.4, pp.217–233 (1998) (in Japanese)

41. Hashimoto, H., Kawasaki, Y., Kobayashi, K., Koike, J., Saito, T., Sugiura, K.: Ecological Cultivation Ark (ECA) Project -Mutation and Evolution of Microorganisms in Space-, *Biological Sciences in Space*, Vol.12, No.2, pp.112–114 (1998)

42. Sugiura, K.: A Materially-closed Aquatic Ecosystem: A Useful Tool for Determining Changes of Ecological Processes in Space, *Biological Sciences in Space*, Vol.12, No.2, pp.115–118 (1998)

43. Sugiura, K.:Ecological Toxicity Testing: Statistical Extrapolation Models and Ecosystem Models, *Japanese Journal of Environmental Toxicology*, Vol.1, No.2, pp.41–53 (1998) (in Japanese)

44. Inamori, Y. and murakami, K.: Effectiveness of Bioassay as Environmental Indicator, *Chemical Industry*, Vol.50, No.9, pp.657–667 (1999) (in Japanese)

45. Sugiura, K.: Effects of Al^{3+} Ions and Cu^{2+} Ions on the Microcosms with Three Different Biological Complexities, *Journal of Sagami Women's University*, Vol.63, pp.53–65 (1999) (in Japanese)

46. Sugiura, K., Hashimoto, H., Ishikawa, Y., Kawasaki, Y., Kobayashi, K., Seki, K, Koike, J., Saito, T.:Cultivation of Bacteria with Ecological Capsules in Space, *Advanced Space Research*, Vol.23, No.2, pp.405–408 (1999)

47. Inamori, Y. and Murakami, K.: Safety Assessment of Foreign Microorganism Introduction using Microcosm, *Environmental Conservation Engineering*, Vol.30, No.6, pp.426–434 (2001) (in Japanese)

48. Sugiura, K.: Effects of Al^{3+} Ions and Cu^{2+} Ions on Microcosms with Three Different Biological Complexities, *Aquatic Toxicology*, Vol.51, pp.405–417 (2001)

49. Sugiura, K., Ishikawa, Y., Shoji, N., Ankyoi, K., Kitazawa, H., Kinoshita, M., Murakami, A., Yoshida, H. and Kawasaki, Y.: A Mathematical Model for Microcosms: Formation of the Colonies and Coupled Oscillation in Population Densities of Bacteria, *Journal of Sagami Women's University*, Vol.66B, pp.27–49 (2002) (in Japanese)

50. Sugiura, K., Kawasaki, Y., Kinoshita, M., Murakami, A., Yoshida, H., Ishikawa, Y.: A Mathematical Model for Microcosm: Formation of Colonies and Coupled Oscillation in Population Densities of Bacteria, *Ecological Modeling*, Vol.168, pp.173–201 (2003)

51. Ishikawa, Y., Yoshida, H., Kinoshita, M., Murakami, A., Sugiura, K.: Examination of the Smallest CELSS (microcosm) through an Individual-based Model Simulation, *Advances in Space Research*, Vol.34, pp.1517–1527 (2004)

52. Murakami, A., Kinoshita, M., Ishikawa, Y., Yoshida, H. and Sugiura, K.: A Mathematical Modeling for Microcosms (Part 1) : Process of Colony Formation, *Eco-Engineering*, Vol.16, No.2, pp.161–170 (2004) (in Japanese)

53. Murakami, A., Kinosita, M., Ishikawa, Y., Yoshida, H. and Sugiura, K.: A Mathematical Modeling for Microcosms (Part 2) : Conditions of Forming Colonies, Stability and Efficiency of the Microcosm, *Eco-Engineering*, Vol.16, No.2, pp.171–180 (2004) (in Japanese)

54. Sugiura, K.:Effects of Chemicals and Metals on Microcosms: Comparison with the NOECs in Experimental and Natural Ecosystems, *Journal of Sagami Women's University*, Vol.68B, pp.1–9 (2004) (in Japanese)

55. Nakane, M., Oumaru, T., Ishikawa, Y. and Sugiura, K.:A Mathematical Modeling for Microcosms (Part 3) : System Diversity and its Stability, *Eco-Engineering*, Vol.20, No.1, pp.19–25 (2008) (in Japanese)

56. Sugiura, K.: Effects of Chemicals and Metal Ions on Microcosms: Comparison of Community Metabolism to Single Species Responses to Toxicants, *Japanese Journal of Environmental Toxicology*, Vol.12, pp.41–53 (2009)

57. Oumaru, T., Nakane, M., Ishikawa, Y., Sugiura, K.: Phase Transition in the Relation between Consumer's Multiplication Ratio and System's Entropy Production in a Microcosm, *Eco-Engineering*, Vol.21, No.2, pp.75–79 (2009) (in Japanese)

58. Ishikawa, Y., Sugiura, K., Nakane, M. and Oumaru, T.: Self-organization for Coexistence in Ecosystem, *Journal of the Japan Association for Philosophy of Science*, Vol.36, No.2, pp.59–66 (2009) (in Japanese)

59. Sigiura, K.; Effects of Chemicals and Metal Ions on Microcosms: Comparison of Community Metabolism to Single Species Response to Toxicants, *Japanese Journal of Environmental Toxicology*, Vol.12, No.1, pp.41–53 (2009)

60. Sugiura, K.: Effects of Chemicals on Microcosms: Comparison with the NOECs in Experimental and Natural Ecosystems, *Japanese Journal of Environmental Toxicology*, Vol.13, pp.13–22 (2010)

61. Fuma, S., Ishii, N., Takeda, H., Miyamoto, K., Yanagisawa, K., Doi, K., Kawaguchi, I., Tanaka, N., Inamori, Y. and Polikarpov, G.G.: Effects of acute γ-irradiation on the aquatic microbial microcosm in comparison with chemicals, *Journal of Environmental Radioactivity*, Vol.100, pp.1027–1033 (2010)

62. Fuma,S., Ishii, N., Takeda, H., Doi, K., Kawaguchi, I., Shikano, S., Tanaka, N. and Inamori, Y.: Effects of Acute γ-Irradiation on Community Structure of the Aquatic Microbial Microcosm, *Journal of Environmental Radioactivity*, Vol.101, pp. 915–922 (2010)

63. Murakami, K.: Ecological Impact Assessment of Metals on P/R Ratio in Experimental Microcosm System, *Papers on Environmental Information Science*, Vol.24, pp.399–404 (2010) (in Japanese)

64. Murakami, K., Hayashi, H. and Shimada, R.: Microcosm for Impact Risk Assessment of Sediment Remediation Materials on Aquatic Ecosystem, *Journal of Water and Environmental Technology*, Vol.9, No.4, pp.401–410 (2011).

65. Inamori, Y. and Fuma, S.: Ecological Assessment of Radiation using Microcosm, *Journal of Water and Waste*, Vol.53, No.7, pp.5–8 (2011) (in Japanese)

66. Murakami, K. and Hayashi, H.: Establishment of Experimental Standardized Microcosm System from the viewpoint of Stability and Reproducibility, *Papers on Environmental Information Science*, Vol.25, pp.221–226 (2011) (in Japanese)

67. Murakami, K. and Hayashi, H.: Effect of Organic Loading in Different Trophic Level on Experimental Microcosm System, *Report of Chiba Institute of Technology*, Vol.58, pp.11–17 (2011) (in Japanese)

68. Fuma, S., Kawaguchi, I., Kubota, Y., Yoshida, S., Kawabata, Z., Polikarpov, G. G.: Effects of Chronic Gamma-Irradiation on the Aquatic Microbial Microcosm: Equi-Dosimetric Comparison with Effects of Heavy Metals, *Journal of Environmental Radioactivity*, Vol.104, pp.81–86 (2012)

69. Hayashi, H., Murakami, K. and Inoue-Kohama, A: Analysis of Ecosystem Impact Statement for Biomanipulation Using the Microcosm, *Journal of Japan Society on Civil Engineering G*, Vol.68, No.7, pp.III_635-III_640 (2012) (in Japanese)

70. Hayashi, H. and Murakami, K.: Ecological Impact Assessment of the Biomanipulation by the Predator Introduction in Microcosm, *Papers on Environmental Information Science*, Vol.26, pp.357–362 (2012) (in Japanese)

71. Kakazu, K., Usui, H., Kumada, J., Ugiura, K., Inamori, R., Inamori, Y.: Ecotoxicity Assessment of SDS with P, R (production, respiration) about Aquatic Microcosm, *Journal of Water and Waste*, Vol.54, No.9, pp.683–690 (2012) (in Japanese)

72. Murakami, K. and Hayashi, H.: Development of Experimental Microcosm System for Environmental Risk Assessment using P/R Ratio Analysis, *Report of Chiba Institute of Technology*, Vol.59, pp.27–31 (2012) (in Japanese)

73. Lyu, Z, Kakazu, K., Sugiura, N., Inamori, R., Xu, K, Murakami, K. and Inamori, Y.: Environmental Impact Risk Assessment of Surfactant Alcohol Ethoxylates using Experimental Model Ecosystem with Aquatic Animals and Plants, *Japanese Journal of Water Treatment Biology*, Vol.49, No.1, pp.21–30 (2013) (in Japanese)

74. Terao, T., Sugiura, K., Nakane, M. and Ishikawa, Y.: Influence of Changes in Growth Rate by Chemicals on Material Circulation Phase and Entropy Production Rate of Microcosm, *Eco-Engineering*, Vol.25, No.4, pp.101–109 (2013) (in Japanese)

75. Hayashi, H. and Murakami, K. : Ecosystem Risk Assessment of Biomanipulation using Experimental model Ecosystem, *Report of Chiba Institute of Technology*, Vol.60, pp.57–62 (2013) (in Japanese)

76. Murakami, K. and Gomyo, M.: Environmental Assessment of Nonradioactive CsCl using Experimental Model Ecosystem, *Papers on Environmental Information Science,*, Vol.28, pp.155–160 (2014) (in Japanese)

77. Terao, T., Nakane, M., Ishikawa, Y. and Sugiura, K.: Space Effect on Microcosm: Analysis of Two Decomposers Coexistence and Biomass Oscillations by Individual-Based Model, *Eco-Engineering*, Vol.26, No.2, pp.35–44 (2014) (in Japanese)

78. Kakazu, K., Inamori, R., Xu, K. and Inamori, Y.: Batch and Repetitive Exposure Assessment of SDS by Indicator Production and Respiration in Aquatic Microcosm, *Journal of Bioindustrial Science*, in printing(2014.2.7 accepted)

79. Shibata, K., Amemiya, T. and Itoh, K.: Effects of Oxytetracycline on Populations and Community Metabolism of an Aquatic Microcosm, *Ecological Research*. DOI 10.1007/s11284-014-1128-3 (2014)

80. Murakami, K.: Environmental Impact Risk Assessment of Microbial Pesticides using Microcosm System, *Journal of GEOMATE* (2018 in submission)

Literature Cited

Bennet A, Bogorad L. Complementary chromatic adaptation in a filamentous blue-green alga. J Cell Biol. 1973;58:419–35.

Beyers RJ. The metabolism of twelve aquatic laboratory microecosystem. Ecol Monogr. 1963;33:281–306.

Beyers RJ, Odum HT. Ecological microcosms. New York: Springer; 1993. 557pp

Burnison BK. Modified dimethyl sulfoxide (DMSO) extraction for chlorophyll analysis of phytoplankton. Can J Fish Aquat Sci. 1980;37:729–33.

Graney RL, Kennedy JH, Rodjers JH Jr. Aquatic mesocosm studies in ecological risk assessment. Boca Raton: Lewis Publishers; 1994. 723pp

Harris CC. Microcosms; ecology, biological implications and environmental impact. New York: Nova Publishers; 2013. 189pp

Holt JG. Bergey's manual of determinative bacteriology. 9th ed. Baltimore: Lippincott Williams & Wilkins; 1994.

Ichihara K. Statistics of the bioscience – practice theory for right and inflects. Nankoudo; 1990, 378pp

Japan Sewage Works Association. Wastewater Examination Method-FY2012-(2012).

Japanese Red Cross Toyota Nursing University Extension. 2005 "basics of statistical analysis to make use of in nursing"; 2005.

Jorgensen SE. Modeling for fate and effect of toxic substances in the environment, Developments in Environmental Modelling 6. Elsevier; 1984. 342pp.

Kakidume T. About the dose-response-related examination in the clinical trial. Lecture Note of Institute of Mathematics Analysis. 2008;1603:1–10.

Kataya A, Matsufuji M. Introduction to Environmental Statistics -Viewpoint, summary of environment data, Ohmsha; 2003. 164pp.

Kurihara Y. Studies of succession in a microcosm, science report of Tohoku University, 4 th Ser. (Biology) 1978; 37:151–160.

Kurihara Y. Studies of the interaction in a microcosm, science report of Tohoku University, 4th Ser. (Biology) 1978; 37:161–177.

Landner L. Chemicals in the aquatic environment, advanced Hazard assessment: Springer. 415pp; 1989.

Leffler JW. The use of self-selected, generic aquatic microcosms for pollution effects assessment. In: White HH, editor. Concepts in marine pollution measurement. College Park: University of Maryland; 1984. p. 139–58.

Matsuda H. Ecological risk science for beginners: precautionary adaptive management. Kyoritsu Shuppan; 2008. 213pp.

© Springer Nature Singapore Pte Ltd. 2020

Y. Inamori (ed.), *Microcosm Manual for Environmental Impact Risk Assessment*,
https://doi.org/10.1007/978-981-13-6798-4

Neison G, Hairston SR. Ecological experiments –purpose, design, and execution: Cambridge University Press; 1989.

Ono S. The basics of analysis of variance to understand by reading. 2nd ed; 2003.

Patel A, et al. Purification and characterization of C-Phycocyanin from cyanobacterial species of marine and freshwater habitat. Protein Expr Purif. 2005;40:248–55.

Peterson JE, Kennedy VS, Dennison WC, Michael Kemp W. Enclosed experimental ecosystems and scale; tools for understanding and managing coastal ecosystems. Springer; 2009. 221pp.

Resetatits Jr. WJ, Bernoado J. Experimental ecology, issues and perspectives. Oxford University Press; 1998. 470pp.

Saijo Y, Sakamoto M. Experimental Analysis of Lake Ecosystem using Mesocosms. Nagoya University Press; 1993. 346pp.

Shiga Scientific Teaching Materials Research Committee. Japanese freshwater plankton illustrated handbook, GodoShuppan;2005. 150pp.. (in Japanese)

Sudo R, Inamori Y. Diagnosis of water treatment from microbiota. 287pp ed: The Industrial Water Institute; 1997. (in Japanese)

Suzuki M, Utsumi H. Bioassay – risk management of water environment, Kodansha Scientific; 1998. 270pp.

Tanaka M. An Illustrated Reference Book of Japanese Freshwater Plankton, The University of Nagoya Press; 2002. 584pp. (in Japanese)

Tanaka N, Inamori Y, Murakami K, Akamatsu T, Kurihara Y. Effect of species composition on stability and reproductivity of small-scale microcosm system. Water Sci Technol. 1994;30 (10):125–31.

Tatarazako, N. The newest domestic and overseas trend of assessment and regulation of effluent water using bioassay – overseas operation case to Japanese WET Test Introduction-, NTS; 2014. 322pp.

Taub FB. Measurement of pollution in standardized aquatic microcosms. In: White HH, editor. Concept in marine pollution measurements. College Park: University of Maryland; 1984. p. 139–58.

The Japanese Society of Environmental Toxicology. Handbook of ecosystem assessment test – environmental risk assessment of chemicals, Asakura Shoten; 2003. 349pp.

The Japanese Society of Environmental Toxicology. Ecological risk assessment and regulation of chemical substances – agricultural chemicals-, Industrial Publishing & Consulting, Inc; 2006. 366pp.

Urano K, Matsuda H. Principles and methods for ecosystem risk management. Ohmsha; 2007. 209pp.

U.S. Environmental Protection Agency. Microcosms as potential screening tools for evaluating transport and effects of effects of toxic substances; 1980. 379pp.

U.S. Environmental Protection Agency. Microcosms as test systems for the ecological effects of toxic substances: an appraisal with cadmium, Athens, U.S. Environmental Protection Agency 1981. 171pp.

U.S. Environmental Agency. Community structure, nutrient dynamics, and the degradation of diethyl phthalate in aquatic laboratory microcosms; 1982. 135pp.

Utsumi H, Nakamuro K. Bioassay and bio-informatics, for environmental assessment and medicinal sciences. Tokyo 214pp: Kougaku-Tosho, Publishers, Ltd; 2011.

Wakabayashi A. Chemicals and ecotoxicity, Japan Environmental Management Association for Industry; 2000. 486pp.

Wellburn AR. The spectral determination of chlorophylls a and b, as well as total carotenoids, using various solvents with spectrophotometers of different resolution. J Plant Physiol. 1994;144:307–13.

WHO. Principles for evaluating health risks to reproduction associated with exposure to chemicals, Environmental Health Criteria 225. Geneva: WHO; 2001.

Yoshida K, Nakanishi J. Basic environmental risk analysis. Tokyotosho; 2006. 243pp

Index

© Springer Nature Singapore Pte Ltd. 2020
Y. Inamori (ed.), *Microcosm Manual for Environmental Impact Risk Assessment*,
https://doi.org/10.1007/978-981-13-6798-4